SKEPTOID 2
MORE CRITICAL ANALYSIS OF POP PHENOMENA

BY BRIAN DUNNING

FOREWORD BY MICHAEL SHERMER
ILLUSTRATIONS BY NATHAN BEBB

Skeptoid 2: More Critical Analysis of Pop Phenomena
Copyright © 2008 by Brian Dunning
All Rights Reserved.

Skeptoid Podcast © 2008 by Brian Dunning
http://skeptoid.com

Published by Skeptoid Media, Inc.

ISBN: 978-1440422850
Printed in the United States of America

We have arranged things so that almost no one understands science and technology. This is a prescription for disaster. We might get away with it for a while, but sooner or later this combustible mixture of ignorance and power is going to blow up in our faces.

Carl Sagan

Contents

Foreword: How to Be a Skeptologist	1
Introduction: Uphill through Mud	5
1. Inside the World's Most Haunted House	7
2. Science Magazines Violating Their Own Missions	13
3. The Twin Towers: Fire Melting Steel	16
4. Mercury, Chelation, and Autism: A Recipe for Risk	20
5. Bizarre Places I'd Like to Go	25
6. Email Myths	30
7. Fluoridation: Death from the Faucet!	36
8. Who Are the Raëlians, and Why Are They Naked?	41
9. Will Drinking from Plastic Bottles Kill You?	46
10. Irradiation: Is Your Food Toxic?	50
11. Crop Circle Jerks	54
12. Subliminal Seduction	60
13. The Attack of Spring Heeled Jack	65
14. How to Argue with a Young Earth Creationist	69
15. The Greatest Secret of Nostradamus	76
16. Do Your Body Features Measure Up?	83
17. Ann Coulter, Scientist	88
18. Raging (Bioidentical) Hormones	92
19. How to Drink Gnarly Breast Milk	96
20. Electromagnetic Hypersensitivity: Real or Imagined?	100
21. A Magical Journey through the Land of Logical Fallacies	107
22. How to See Your Aura	120
23. Who Kills More: Religion or Atheism?	125
24. Orang Pendek: Forest Hobbit of Sumatra	129

25. Medical Myths in Movies and Culture	134
26. Aliens in Roswell	138
27. Death in Your Kitchen: Microwave Ovens	145
28. Ghost Hunting Tools of the Trade	151
29. What Do Creationists Really Believe?	156
30. The Detoxification Myth	162
31. Magic Jewelry	167
32. World Trade Center 7: The Lies Come Crashing Down	172
33. MonaVie and Other "Superfruit" Juices	177
34. Water as an Alternative Fuel	182
35. Super-Sized Fast Food Phobia	187
36. Despicable Vulture Scumbags	192
37. Can You Hear The Hum?	197
38. The "Terror" of Nuclear Power	203
39. Apocalypse 2012	208
40. Fire in the Sky: A Real UFO Abduction?	213
41. Bend Over and Own Your Own Business	218
42. What's Wrong with The Secret?	223
43. The Face on Mars Revealed	228
44. The Crystal Skull: Mystical, or Modern?	232
45. Reassembling TWA Flight 800	236
46. Is Peak Oil the End of Civilization?	242
47. What You Didn't Know about The Stanford Prison Experiment	247
48. Should You Take Your Vitamins?	252
49. When People Talk Backwards	257
50. King Tut's Curse!	261

Foreword: How to Be a Skeptologist

by Michael Shermer

I am not a psychic, but as a professional skeptic I occasionally play one to reveal the tricks used by peddlers of the paranormal. The most common ruse is known as cold reading, where you reveal facts about someone you have never met. It is not difficult. Armed with the knowledge that certain facts are likely to apply to anyone (e.g., a scar on your knee, a white car in your past, the number two in your address), a friendly and confident patter punctuated with inquisitive looks and knowing nods — and no moral scruples — you too can be a psychic, astrologer, palm reader or tarot card diviner. No matter how you market yourself, the process is the same.

And it's easy to find customers, because a great number of people are ready to believe. According to a 2005 Gallup poll, three-quarters of those surveyed believe in at least one paranormal phenomenon, including 41 percent who are convinced of ESP, 32 percent of ghosts, 31 percent of mind reading, 26 percent of clairvoyance and 25 percent of astrology.

Spend 10 minutes online and you can catalog many other highly questionable beliefs that aren't related to the paranormal, such as that space aliens landed at Roswell, New Mexico, that the earth was created less than 10,000 years ago, that the Holocaust never occurred, and that 9/11 was orchestrated by the U.S. government to galvanize America for war.

Why do so many people believe such weird things?

First, all humans seek patterns. That's our nature. We are also storytellers because it helps us find meaning in a chaotic world. In order to survive, we have evolved to find cause-and-effect relationships in nature, and then weave a plausible story to explain them. Our ancestors who identified the pattern linking the seasons to animal migrations ate better and left behind more offspring. But because believing that the rain gods can be appeased through rituals isn't fatal, we also have inherited magical thinking. Add to this the fact that many of these beliefs make us feel better, meet some emotional need, promise miracle cures or instant wealth, and in general appeal to our emotional brains and bypass our rational brains.

What can we do about this? Think skeptically. How? Here are a few questions to ask when considering extraordinary claims:

1. Is the person making this claim a qualified expert in the field, or a quack?

People who are not trained in a subject can make contributions, but it is rare.

2. Does the source often make similar claims?

Paranormalists and members of fringe groups have a habit of going well beyond the facts.

3. Have the claims been verified by another source?

Typically pseudoscientists will make statements that are unverified, or verified by a source within their own circle. Who is checking the claim, and who is checking the checkers?

4. How does the claim fit with what we know about how the world works?

When considered in this manner, get-rich quick schemes and stock-market secrets never sound so good.

5. Has anyone gone out of the way to disprove the claim, or have they only sound evidence to confirm it?

This is known as confirmation bias, or the tendency to ignore negative evidence. This is why we need the methods of science, which include the attempt to prove yourself wrong.

6. Does the preponderance of evidence converge to the claimant's conclusion?

The theory of evolution, for example, is proven through a convergence of evidence from a number of independent lines of inquiry. No one fossil proves anything.

7. Is the claimant employing accepted rules of reason and tools of research?

UFOlogists suffer this fallacy in their continued focus on a handful of unexplained atmospheric anomalies and visual misperceptions while ignoring the fact that the vast majority of sightings are easily explained.

8. Has the claimant provided a different explanation for the observed phenomena, or is it strictly a process of denying the existing explanation?

This is a classic debate strategy—criticize your opponent and never affirm what you believe in order to avoid criticism. Creationists do this to great effect. But to be legitimate, positive evidence in favor of your idea must also be presented.

9. If the claimant has offered a new explanation, does it account for as many phenomena as the old explanation?

For example, skeptics who argue that lifestyle, not HIV, causes AIDS do not explain nearly as much of the data as the HIV theory does, such as the rise in AIDS among hemophiliacs shortly after HIV was inadvertently introduced into the blood supply.

10. Is there extraordinary evidence for the extraordinary claim?

Evidence is key. Normal claims need normal evidence, but extraordinary claims require extraordinary evidence.

But this is just a start. There is, today, a considerable body of skeptical literature outlining how to think like a skeptic. One of the very best is Brian Dunning's Skeptoid II. Flip open to any page in this exceptionally readable and highly informative book by one the America's top skeptical investigators and bloggers and you won't be able to stop reading. Jumping from one topic to another, Dunning reveals the depth of his critical faculties and the breadth of his wide-ranging interests, and yet throughout the book one central theme comes through: evidence and logic is all that matters when evaluating a claim. It doesn't matter what your name is or what degrees you have or how emotionally compelling your arguments are, Dunning unfailingly demands evidence for your claim and applies logic to his analysis. If you can convince Dunning then you've jumped a mighty skeptical hurdle on the way to truth.

— Michael Shermer

Michael Shermer is the Publisher of Skeptic magazine, monthly columnist for Scientific American, and the author of Why People Believe Weird Things, How We Believe, Why Darwin Matters, and The Mind of the Market.

Introduction: Uphill through Mud

Almost nobody wants to hear what skeptics have to say.

Nobody wants to hear that a vitamin pill won't cure their cold, that financial independence is more than just a few books and tapes away, or that physical fitness requires more work than slipping on an ionized bracelet.

People want to be sold easy answers. They want to know that everything is in reach and is simple. People want superpowers, so they want to believe that it's possible to speak to dead people, to see into the future, to read minds, to have a super duper immune system, or that alien races visit one another. Offer to sell those easy answers, and you'll have customers lined up at your door, fistfuls of cash at the ready.

But offer to protect their wallets against such nonsense, and nobody wants to listen. The skeptical perspective is not the one that sells. No late night infomercial companies process millions of dollars in orders for books and tapes promising only cold hard reality.

And that's a shame, because reality is actually the secret to success. A decision well informed with facts is always more useful than a decision based on superstition. The patient who treats his diabetes with insulin will nearly always outlive the patient who relies on an untested herbal extract. The small businessman who works hard to provide a good service at a fair price will nearly always grow wealthier than someone who buys into a multilevel marketing pyramid scheme. The driver who buys an efficient car and drives cautiously will nearly always get better mileage than someone who buys implausible, scientific-sounding products off the Internet guaranteed to "double your mileage". A parent who understands perceptual phenomena and psychology is far better prepared to comfort a frightened child than a parent who believes there really is a ghost in the closet. The ability to apply skepticism and think critically is the single most valuable ability a person can have, and the most universally applicable to all facets of life.

Everyone is a skeptic to one degree or another. Even the most hardcore, dyed in the wool believer in the supernatural is skeptical about something. If you're reading this book, think of some popular urban legend or myth that you think is silly: Bigfoot, alien abductions, the Loch Ness Monster, the Bermuda Triangle. Why is it silly? What about it is implausible? What's a better explanation for the reported claims?

That's skepticism. By applying critical thinking, you've just arrived at (probably) a better explanation that is more consistent with what we've learned about our world and our universe. If you ever need to make a decision about that subject, you're better equipped than some other people to make a wise and informed decision.

An experienced skeptic is much less likely to be taken advantage of: Critical thinkers are accustomed to the patterns that typically identify pseudoscience and frauds. Skeptics are more likely to discover scientific advances: An appreciation for and understanding of the scientific method, including the ability to discriminate between well sourced evidence and poorly sourced evidence, is a crucial element of scientific progress. A skeptic is much less likely to waste his money on useless alternative therapies, and instead banks his health on evidence based medicine.

These are the kind of benefits that promoters of critical thinking are trying to push on you, and that nobody wants to hear. I find this ironic, because I discover excitement and fascination every day in the process of illuminating facts from the darkness of fallacy.

It is in the hope of bringing this illumination to others that I continue trudging uphill through mud.

— Brian Dunning

1. Inside the World's Most Haunted House

Gather now as we throw another log on the fire, pour some milk in our tea, and close the shutters against the mist as we tell stories of Borley Rectory, the most haunted house in England, and probably in the world.

A rectory is the residence provided by a church to its rector, vicar, or minister. This particular rectory was built on the same site as a Cistercian priory perhaps several hundred years older in what is now Borley, Essex, United Kingdom. There are two stories of ancient love affairs gone wrong from Borley Rectory. In one account, a monk from a nearby 14th century monastery had a relationship with a novice from the local nunnery at Bures. When the illegal affair was discovered, the monk was hanged and the nun was bricked up alive inside the basement of the priory, which later became Borley Rectory. Later, in the 17th century, a French nun named Marie Lairre left her order in Le Havre and came to England, staying for some time at the same nunnery in Bures. Soon she met and married Henry Waldengrave, owner of a manor home that stood on the site of Borley Rectory. In an evening of rage, Waldengrave strangled his wife, and buried her in the basement.

Eventually, in 1862, Borley Rectory was constructed for the Reverend Henry Dawson Ellis Bull. Almost from the beginning, the Bull family was plagued by frightening apparitions. A ghostly nun was frequently reported in the twilight near the home, walking through the gardens. Once Bull's daughters tried to talk to the nun, only to see her fade away and disappear as they got closer. The family was shocked to learn that the nun's path through the garden was already well known to the local villagers, and was called the "Nun's Walk". Sometimes the nun was seen watching people from an upstairs window. Even more terrifying was the appearance of a phantom coach driven by two headless coachmen, which was sometimes seen and often heard at night in front of the rectory. The sounds of mysterious footsteps and strange creaks and crashes were commonly heard inside the house. Reverend Bull's son, Harry Bull, succeeded his father and stayed in the home until his death in 1927. It was said that Harry Bull enjoyed the ghostly disturbances as entertainment, and built a summer house overlooking the Nun's Walk where he could enjoy cigars and watch the spectacle.

The new rector, Guy Smith, moved in with his family in 1928. While cleaning, Mrs. Smith found a strange package wrapped in brown paper, and inside was the skull of a young woman. The same strange incidents plagued the Smith family, and after Mrs. Smith saw the phantom coach, they called in *The Daily Mirror* newspaper for help. *The Daily Mirror* sent paranormal researcher Harry Price to investigate. Price had stones and a vase thrown at him from unseen hands. After the Smith's daughter was inexplicably locked in a room with no key, they had enough, and moved out after only one year.

The next victims were Reverend Lionel Foyster and his wife Marianne, and it was during their stay that Borley Rectory's most famous haunting occurred: the appearance of automatic writing on the walls of the house. The writings contained pleas for help from Marie Lairre, often addressed specifically to Marianne. The writings said things like "Marianne, please help get" and "Marianne, light mass prayers" and "Pleas for help and prayers". The writings sometimes even appeared in real time while people watched! The Foysters tried to erase and even paint over the writing, but it persisted.

Marianne was often victimized by violence. She was thrown from her bed on many occasions, was attacked and slapped by unseen assailants, and was struck by flying rocks. Windows shattered spontaneously. Reverend Foyster tried many times to exorcise the house, without result, and kept logs of the incidents which he mailed to Harry Price. Price said that the Foysters reported as many as 2,000 events.

The Foysters finally gave up and moved out, and Harry Price himself rented Borley Rectory. Price advertised for 48 volunteer researchers to come and stay in the house with him and help record the supernatural episodes. Along with his best friends and fellow researchers Sidney and Helen Glanville, Price conducted seances using a planchette, a writing implement held by the seance participants similar to a Ouija Board. Two spirits most often manifested themselves during these seances. Marie Lairre, the most vocal of the spirits, told her woeful story and explained that she was condemned to wander until her bones could receive a proper Christian burial. The second spirit, named Sunex Amures, warned that he would burn down the rectory that very night, and that the bones of a murder victim would be revealed in the wreckage.

The rectory did burn down, but it was eleven months after the ghostly threat. The home's new owner, Captain W. H. Gregson, was unpacking and accidentally overturned an oil lamp, starting a fire that destroyed the building. During the inferno, onlookers spotted a nun in one of the windows. Afterward the rubble was demolished and the bricks were re-used for the war effort, leaving a bare hole in the ground.

Harry Price took advantage of the unfortunate opportunity and excavated the basement. The bones of a young woman were found, certified by a pathologist, and reburied in the nearby cemetery at Liston in 1943. After nearly a century of haunting, Marie Lairre was finally at rest, the Nun's Walk found peace, and the legend of the most haunted house in England came to an end.

And now, decades later, we turn a skeptical eye upon Borley Rectory and see how much of it we can verify, and how much of it is complete bunk. One of the keys to understanding the events at Borley Rectory is to understand who Harry Price was. By no means was he a scientist or an unbiased researcher. He was an expert magician, a member of the British organization The Magic Circle, and proven hoaxer. He was a close friend of Charles Dawson, the man behind the infamous Piltdown Man hoax. He and photographer William Hope staged an elaborate photograph depicting a ghost looking over the shoulder of Price as he sat for a portrait. Harry Price went on the road with a fake statue of Hercules. He exhibited a fake silver ingot from the reign of Roman emperor Honorious. He showed gold coins from the kings of Sussex and a bone carved with hieroglyphics, all proven to be fakes. By every account, Harry Price was a practiced hoaxster and very much of the P. T. Barnum mold. Harry Price did not investigate Borley rectory for his own health. He achieved a great deal of notoriety from it, including the publication of three books, *The Most Haunted House in England*, *Poltergeist Over England*, and *The End of Borley Rectory*.

It's important to note that prior to the 1929 article in *The Daily Mirror*, when Harry Price was first called in, no written account exists of any unusual incidents at Borley Rectory. A closer look at the facts reveals a long string of inconsistencies and contradictions between Price's published accounts and the reports of the families themselves. Let's go through a few of these.

The legend of the nun bricked up in the cellar, that so frightened the Bull family, came from a novel that they owned by Rider Haggard. Reverend Bull used to read this chilling tale to his children.

Reverend and Mrs. Smith said that they left the house due to its horrible condition and prehistoric plumbing, not due to any hauntings. The skull that Mrs. Smith found was attributed to a victim of the 1654 plague, many victims of which were crudely buried in the ground that later became part of the garden of Borley Rectory. It was not uncommon for skulls and other bones to be found on the property, and they were routinely reburied in the churchyard.

Marianne Foyster stated that she believed many of the strange incidents were being staged by her husband working in league with Harry Price. Harry Price countered that he believed Marianne herself was, consciously or unconsciously, causing some of the incidents, stating that events only seem to occur when she was present.

There is much confusion over the automatic writing. Most significantly, accounts of the Glanville's seances show that they used rolls of wallpaper on which to capture the writings of their planchette. Why they used wallpaper rolls is not clear, but it could be as simple as wallpaper being the largest rolls of paper that were handy. The story of automatic writing appearing on the walls of Borley Rectory while people watched appears to be nothing more than a misinterpretation of the reports of the planchette seances, in which writing was captured on wallpaper while seance attendees watched and participated. As for the contents of the writings, most are almost completely illegible, and the popular interpretations are dubious at best. In particular, the writing interpreted as the name Marie Lairre appears to many skeptics to say no such thing.

When Borley Rectory burned down, the insurance company determined the fire to be arson, and Captain Gregson's claim to be fraudulent. What connection this has to Harry Price is not certain, but Gregson was instrumental in organizing Price's excavation, and was present when the bones were found in the cellar. You decide.

Price's discovery of the bones has also been the subject of debate. Critics have questioned the likelihood of Price turning up bones in a single search in only a few hours, when other searches, both before and after Price's excavation, came up empty handed despite far more extensive digging. They also question the fortuitous presence of a pathologist and a barrister to certify the remains. And to make it even more confusing, the two gardeners who did the actual digging, Johnnie Palmer and Mr. Jackson, identified the only bone recovered as a pig's jawbone. What was actually recovered, and how did Price happen to have a pathologist and a barrister on hand? It's unlikely that we'll ever know either answer for sure, but there's enough uncertainty to put Price's own claim on thin ice.

Harry Price died only a few years later, and some of his former associates from the English Society for Psychical Research published their own findings and analysis. A similar report was made by the London Society for Psychical Research. Both reports concluded that (1) there were no verifiable events that could not have had natural explanations, (2) that Harry Price's duplicity made it hopeless to determine the validity of his findings, and (3) that the most popularized events were caused by Harry Price himself. They even debunked specific episodes, such as a light often seen in one of the rectory's upper windows happened to coincide with the reflected headlight of a regularly scheduled train nearby.

The conclusion I draw from all of this is that to enjoy a good ghost story, you'd better not look at it too closely. If the events at the world's most haunted house can be total fabrications, then what about all those other lesser hauntings around the world? Maybe it's time for one of them to step up and take over the crown. All it takes is some creativity and a book with a great title.

2. Science Magazines Violating Their Own Missions

Today we're going to sit back with our favorite science magazine, open a cold beverage, and read outrageous pseudoscience claims. How's that? Have they lost their editorial way? Not quite: They've lost control of their advertising departments.

I've been a reader of popular publications like *Scientific American* and *Popular Science* for many years. I enjoy the articles but I always have to deliberately avoid the last pages where they tend to run advertisements for blatantly pseudoscientific products: aphrodisiacs, herbal supplements, magic jewelry, and the like. Now obviously, *Popular Science* has to make money and advertising is one way they do that. If their hands were completely tied and they tried to be too restrictive about what types of ads they run, they might not make the money they need. Even skeptical readers like me would rather see them in business than out of it, so we should probably allow them the leeway they need to make the money they need. Right? Well, maybe. It's a give and take. The more bogus ads they run, the more it cuts into their credibility. These publications put themselves out there as proponents of scientific advances, and when they publish even third-party materials that run counter to this mission, they're contributing to society's built-in adhesion to the Dark Ages.

I contacted *Popular Science's* advertising department and asked for their advertising guidelines. I wasn't able to get anything on paper, but I did get a verbal instruction that the products they advertise must actually work and must do what the ad says they do. OK, interesting. Let's open *Popular Science* and see if the advertised products are truly evidenced to do what they claim.

Here's an anti-aging supplement on page 90 that will make you look and feel younger, stronger, perform better, and recover faster. Plus it's "lab tested" and "highest rated". Apparently, that's good enough for *Popular Science*.

On another page I find a "trust potion". It's a spray that "compels others to trust you" and "fuels intimacy". A trust potion, *Popular Science!* Shouldn't this be your cover story? Can we get an article about this? Obviously you must agree that it does what it claims.

I actually did find one ad for a useless product that says, as required, that its statements have not been evaluated by the FDA — but only one, and my understanding of the law is that this is required of all ads that make unsupported medical claims. This one's for a "Super Male Pill from Europe". Oh, it's from Europe! We'd better read on. It's an "all natural super sex pill" and it makes some very specific claims for its male enhancement results that are a little too racy for this book. I counted 24 exclamation points in this one advertisement. Generally, exclamation point count is considered the hallmark of responsible reporting. But not to worry, this product does promise that it will not cause "blue vision".

Here's an ad for a pheromone additive for your cologne. It says it was published in a "respected biomedical journal". But they are looking out for you: They advise you to "reject cheap imitations".

Here's a two-page spread with facing ads for competing male enhancement products. One of them is "doctor approved", offers 5 inches of "enhancement", and sells for $120 for a three month supply. But the other is "natural", offers only 1 to 3 inches of "enhancement", and sells for $327 for a six month supply. Even though the latter product doesn't enhance you as much, it's worth so much more because it's natural and not doctor approved. Once again we have excellent proof that anything all-natural is much better than anything doctor approved.

One of the most bizarre ads in *Popular Science* is for a water filter — or something; neither their ads nor their web site is willing to tell you exactly what their product is — that claims that pure water, filtered water, and distilled water are toxic, and that their special water machine (whatever it is) is the only way to get water that doesn't "spread disease". Among the mess on their home page are claims that going to the dentist can give you AIDS, pure water causes serious prostate problems, water is "dead", Internet search engines "lie to push their own hidden agenda in spite of human suffering", and my favorite, "sellers of pure water products are breaking the law by hiding facts from buyers that they need to make an informed buying decision!" (exclamation point). Thank you, *Popular Science* magazine, for alerting us to these dangers.

Scientific American is perhaps not as guilty of spreading this nonsense as *Popular Science*, but their closet is by no means totally clean either. In nearly every issue they run a full page ad for an outrageously priced exercise machine, and it states clearly that four minutes a day on this machine gives you the same benefit as 20 to 45 minutes of running, plus 45 minutes of weight training, plus 20 minutes of stretching, plus it balances your blood sugar (whatever that means), plus it repairs bad backs and shoulders, plus it will make your body look so good that your friends will all buy one too — all in only four minutes a day. Now I don't want that company to sue me, so I'm not going to make a statement like those claims are all blatantly fraudulent, but I find it bizarre that a magazine with standards for the products they advertise could have read over that copy and found it to be acceptable. And by the way, the same ad is also in *Popular Science*.

If you make the decision that your mission is to advance science, it makes no sense to undermine that mission by publishing ads for products that are fraudulent or that make unsupported bogus claims. So I can only assume that *Popular Science* does not have the advancement of science as their mission, or if they do, it's in some kind of "negotiable" status. So, buyer beware, and read with caution. The standards here are basically the same as those for Oprah or Montel.

3. The Twin Towers: Fire Melting Steel

Today we're going to really put the Men in Black under the microscope. And by Men in Black, I mean blacksmiths. You know, those evil government conspirators who expect us to believe that steel can be melted by something that ignites at a far lower temperature. For thousands of years, blacksmiths have been lying to us. They've been telling us that they use coal to melt steel for casting, which, according to a poster on the Skeptalk email discussion list, burns at about 560°F. Fortunately we know better. We don't buy into their lies. We know that steel melts at 2750°F, so we know that these blacksmith shops at local living history museums are all part of the government's master plan of deception. The whole smithing profession and false history was probably invented by the government to prepare us to believe in their biggest lie: That the fires inside the World Trade Center could have brought the towers crashing down.

Conspiracy theorists love to quote retired New York deputy fire chief Vincent Dunn, who said "I have never seen melted steel in a building fire." But they conveniently omit the second half of his sentence: "But I've seen a lot of twisted, warped, bent and sagging steel. What happens is that the steel tries to expand at both ends, but when it can no longer expand, it sags and the surrounding concrete cracks."

One tactic used by conspiracy theorists that has frustrated engineers is their use of a straw man argument, which is where you repeat your opponent's position and carefully reframe it to be weaker and obviously false. Here, the conspiracy theorists have reframed the engineers' position as stating that the World Trade Center fire melted the steel. This is not true, no such claim has been made, as actual melting was neither necessary for the collapse nor possible with the amount of heat that was available.

Let's review the numbers one more time, if you're not already sick of hearing this over the past six years. Steel melts, or liquefies, at 2750°F. Let's take that off the table, because nobody claims that it got that hot, and it wasn't what happened. Jet fuel burns at up to 1500°F. Within about 10 minutes, the jet fuel was exhausted, and the fire then raged among the building itself: its furniture, rugs, curtains, papers, whatever, and temperatures preceding the collapse reached a maximum of 1832°F, according to the National Institute for Standards and Technology's analysis of heat damage to the debris, and as simulated using their computational fluid dynamics model known as the Fire Dynamics Simulator. According to the American Institute of Steel Construction, "Steel loses about 50 percent of its strength at 1100°F, and at 1800°F it is probably less than 10 percent." Even the lowest end of the temperatures inside the fire were way hotter than the hottest temperatures at which the steel trusses could have maintained integrity.

But for the conspiracy theory to work, you have to dismiss any statements made by any official or independent agency, because they could all be part of the conspiracy. The only figures considered reliable are those which differ significantly from official reports. Even expert Rosie O'Donnell told us "It's the first time in history that fire has melted steel."

But then, on April 29, 2007, fire melted steel for the second time in history. A freeway accident occurred in Oakland, California that made us all take a second look. A tanker truck carrying 8,600 gallons of gasoline lost control and crashed on an elevated underpass in the Macarthur Maze, a knot of converging freeway ramps taking cars from the 24, 80, 580, 880, and 980 freeways and funneling them into the San Francisco - Oakland Bay Bridge toll plaza. The fuel exploded into flames and burned fiercely for several hours, but it only took minutes for the span above the flames to collapse and fall onto the span below. The director of Cal Trans, the California state transportation authority, said the heat from the fire had melted the steel girders and bolts that support the concrete roadway. He said "If you have that kind of heat, you're going to have this kind of reaction. We're not surprised this happened."

The massive I-beams built into the structure of the freeway overpasses are far thicker and heavier than the lightweight steel trusses supporting the floors of the World Trade Center. The speedy and graphic nature of this failure demonstrated once and for all how easy it is for heat to soften steel just enough to sag, and that little sag is all it takes for the structure to come apart and then it's Good Night Ladies. In Oakland, these giant beams didn't just sag: they squished like they were made of clay, as you can see if you look up the photos.

Happily, the freeway collapse did have a silver lining. Engineers everywhere breathed a sigh of relief, since this was such a major bitch-slap to the 9/11 conspiracy theorists. Now maybe those nutballs would shut up and go home, right? Maybe even take down their insulting web sites. But is that what happened? Don't bet on it. Remember how the logic of the conspiracy theorist works: Evidence against their theory is really evidence for the conspiracy. Within hours, conspiracy theorist blogs and web sites were charging that the government staged the Oakland freeway collapse in a transparent attempt to bolster the official version of the World Trade Center events.

Three basic arguments have been made alleging the conspiracy. First, it just seems consistent with what an evil government might do. But, like the majority of the 9/11 conspiracy "evidence", appearing consistent with one possibility in addition to others is hardly proof that that one possibility is the true one.

Second, this fire was outdoors, and not insulated within a building. For some reason the conspiracy guys have turned this one completely around, saying that an uncontained outdoor fire traps heat in better than an enclosed fire. This logic is a little too bizarre for this author to attempt to address. This has nothing to do with oxygen availability, which was the only remotely intelligent extrapolation I could make from this, as the World Trade Center fires were fed not only by airliner sized holes in the side of the building, but also by millions of cubic yards of oxygen inside the buildings.

Finally, the conspiracy guys argue that of all the hundreds of thousands of freeway overpasses in the country, how could this accident just happen to occur at one of the busiest interchanges on the busiest bridge in one of the most traffic congested urban areas in the country? If you wanted to deliberately select the most disruptive and highly visible interchange in the country, this is quite possibly the exact one you'd choose. The two spans that were destroyed carry 160,000 cars a day. What are the chances that this is where such an accident would just happen to occur? Next to impossible. Clearly, this location had to be deliberately chosen. The only possible explanation is that the wreck was staged by the government.

It's kind of hard to argue against that kind of logic. So, I say, don't bother. People who are smart enough to know better, and educated enough to understand the physical sciences, and yet still believe the conspiracy theories, are beyond help. They are paranoid delusionals. Don't waste your breath trying to reason them into mental health. And also, don't worry that their fantasies will eventually creep into the history books and infect your children, any more than you should worry that the schools will start teaching the Flat Earth theory. The conspiracy theories are false, so they're unprovable, and all the evidence will always be against them. They're never going to go away, and they're never going to shut up, and as offensive as their paranoid pipe dreams are to civilized people, they have every right to present them and argue their point of view. This is the lesson for your children. Show your children the facts of what happened, and explain why the terrorists did what they did — that's the easy part — and then expand the lesson to the importance of free speech. Better if your children first hear these conspiracy theories within the context of an example of protected free expression of an offensive idea.

That way, your children will be better prepared to visit a blacksmith shop, and know when they're being lied to.

4. Mercury, Chelation, and Autism: A Recipe for Risk

Today we're going to examine yet another case where people are willing to put their own and their children's lives at risk in order to embrace popular pseudoscience. It seems that more and more, people are increasingly concerned with joining a politically correct fad when it offers a simpler explanation than what medical fact offers. In this case, parents of autistic children have, in the absence of a medical cure for their child's condition, turned to alternative medicine and put their children at greater risk by avoiding crucial vaccinations or even causing direct injury with chelation.

In my Skeptoid podcast episode about amalgam dental fillings, I was widely criticized for mentioning chelation therapy as a valid treatment to remove heavy metals from the body. What I said was misinterpreted as support for the popular misuse of chelation, when it's used for non-existent contamination or for so-called "cleansing". Real chelation therapy is used medically, though rarely, because there is such a thing as real heavy metal contamination that is dangerous, most commonly lead poisoning. It usually happens occupationally to people who work with heavy elements and are involved in accidents. Medical chelation takes years and is, at best, only partially successful; and carries plenty risk of its own. Kidney damage is among the most common side effects. Chelation therapy in popular alternative medicine, however, brings only the risk and no possible benefit to the recipient.

So how did we get to a point where wrongly informed parents are turning to chelation to treat their autistic children? It's not all that surprising. Many of the indications of autism first become apparent in children at approximately the same age as vaccinations are given. It naturally follows that some people will thus draw an (invalid) causal relationship. Because they happened about the same time, one must have caused the other.

This is the same logic flaw that leads Oprah guests to proclaim their cancer was cured by some alternative therapy. Of those lucky few individuals whose cancer spontaneously went into remission, many were probably taking some random alternative therapy at the time; and because the remission occurred about the same time as the therapy, they assumed a causal relationship, when in fact none exists.

No parent wants to see anything bad happen to their child. When it does, it's natural to seek some outside cause, someone or something to blame, something that can be attacked and fought back. Popular media has spread the notion that mercury from vaccination causes autism, and this makes a perfect scapegoat. Something to blame, something to fight, some way to protect the child. An easy answer. A clear answer. A chance. Something more tangible than the doctor's vague explanation of the complex causes of autism, and its tragic incurability. It's the perfect opiate for the psychologically tormented parent.

But it does have its costs. In Pennsylvania, the parents of Abubakar Tariq Nadama, a 5-year-old autistic child killed by chelation therapy in 2005, are suing the individuals and companies involved for wrongful death and lack of informed consent. He was being treated with EDTA, which is approved by the FDA for use only after blood tests confirm acute heavy-metal poisoning. The child's blood tests did not reveal any such poisoning. Howard Carpenter, executive director of the Advisory Board on Autism-Related Disorders, said "It was just a matter of time before something like this would happen." Gary Swanson, a psychiatrist who works with autistic children, said "I can't sit there and endorse it as a viable treatment. It's not something published in peer review journals and studies. It's probably a quack kind of medicine."

As previously mentioned, the exact causes for all the various forms of autism are complicated and are not 100% understood, but that doesn't mean that nothing is known or that non-evidence-based alternative therapy might be useful. One of the factors that is known is that heredity is present in 90% of autism cases. It's largely genetic, not environmental. Studies have determined that a few agents such as thalidomide, when present during the first 8 weeks of gestation, can cause the same chromosomal damage found in autism. No rigorous scientific evidence has ever been found that indicates autism can otherwise be caused environmentally, which eliminates all the pop-culture supposed causes like vaccination, food allergies, or mercury poisoning.

Moreover, a 2007 study by Williams, Hersh, Allard, and Sears published in Research in Autism Spectrum Disorders found no significant difference in the levels of mercury found in hair samples between autistic children and their non-autistic siblings. Siblings were used for this study to eliminate other environmental variables as factors. Consumer Health Digest concludes "Autism has no plausible association with mercury toxicity or other heavy metal exposure."

Proponents of the alleged link between vaccines and autism charge that vaccines contain mercury, which in large enough doses, kills cells and causes neurological damage. What some vaccines contain is actually not plain mercury, but the preservative thimerosal. Thimerosal's main active ingredient is an organic version of mercury called ethylmercury. Ethylmercury is naturally expelled from the body quickly. Methylmercury, on the other hand, is not. It stays in the body. High doses of methylmercury will cause physiological damage. However, ethylmercury and methylmercury are not the same thing, despite the similar names. Methylmercury is not present in thimerosal. In short, vaccines preserved with thimerosal do not even contain the type of mercury that activists say is dangerous. And even if they did, the amount would be too small to be considered a risk.

It doesn't help that this misinformation is spread by celebrity activists like Robert Kennedy Jr., whose only medical experience comes from carefully making lines of cocaine with a razor blade. Kennedy wrote an article for Rolling Stone magazine in 2005 charging that the government knows that vaccines cause autism and is actively covering it up. I wonder what young Abubakar's parents think of Kennedy's contribution to pop-culture. The online version of Kennedy's article is followed by *five paragraphs* of corrections and clarifications, among them pointing out that he misstated the amount of ethylmercury received by infants at six months of age, by a factor of 133 times the actual amount. His article is bursting at the seams with flawed logic and irrelevant comparisons, such as this one (the italics are mine): "Infants routinely received three inoculations that contained a total of 62.5 micrograms of *ethylmercury* – a level 99 times greater than the EPA's limit for daily exposure to *methylmercury*." It's OK though, Robert, people don't read too closely.

Rates of vaccination have not been increasing, so why the reported skyrocketing rates of autism diagnoses? An increasingly broad array of conditions being called autism is part of the reason. Autism is not necessarily a single, well-defined disorder. There are five main Autism Spectrum Disorders, including but not limited to Asperger syndrome, Rett syndrome, various childhood disintegrative disorders, and pervasive developmental disorder not otherwise specified, or PDD-NOS. As more of these are broadly called "autism", obviously the rates of autism rise substantially. Between 1987 and 1998, the number of patients classified as autistic rose 273 percent.

If thimerosal were a cause of autism, then wouldn't its removal from vaccines curb the rising rates of diagnosis? Well, obviously, yes it would. But it didn't. The FDA removed thimerosal from childhood vaccines in the US in 1997, as a precautionary measure, partly in response to all the anti-vaccine activism. Autism diagnoses continued to rise unabated. Denmark and Sweden eliminated thimerosal five years earlier. Their rates also continued to climb, as their definition of autism was also broadened.

Let's repeat that, since apparently it's not clear to Kennedy and the other activists still warning against vaccination. Ethylmercury-containing thimerosal was removed from childhood vaccines in 1997. Vaccination will not result in mercury poisoning.

Vaccinations save more lives worldwide than any other medical advance in history. Thanks to vaccination, children around the world are now safe from hepatitis A and B, polio, smallpox, measles, rubella, diphtheria, tetanus, rotavirus, mumps, typhoid, and many more. Giving up all of these immunities, due only to an unfounded fear of a compound that's no longer used and was demonstrated safe in every rigorous study ever done, is hardly the best way to serve your child. Exposing an already-vaccinated child to the dangers of chelation in a misguided effort to remove nonexistent, undetected poisons is just as bad. Vaccinate your children. Don't put them or yourself through the risks of chelation therapy, unless of course your job at Three-Mile Island was to drink all the leaked cooling water.

5. Bizarre Places I'd Like to Go

Unfold your map of California, boot up your GPS, and hold on tight for a ride around some of our state's most intriguing mystery spots. California is such a big state with such diverse geography and history that we have as many natural oddities as we have ghosts and monsters and mystery lights and just about anything else from the world of the strange. A lot of it is just legend but there's also plenty that's real enough for a memorable weekend road trip. Today I'd like to share with you some of my favorite places in California that either tickle the skeptical funnybone, or are just too darn cool scientifically to pass up.

Some of these places are so unique and valuable that their exact locations are secret. For example, the oldest living thing, a 4700-year-old bristlecone pine named Methuselah, is in California's White Mountains, and its location has been kept a secret ever since an even older tree named Prometheus was cut down in 1964 in order to measure its age. Sort of a Schroedinger's Cat experiment taken to the extreme.

Several times I've visited the most impressive petroglyphs I've ever seen in Death Valley. They're in a secret slot canyon with a name that describes its sharply twisting shape. These huge petroglyphs are amazing and they're absolutely pristine. They're also difficult and dangerous to reach, and impossible when water is flowing through the canyon. Don't expect any help from the rangers either; they've "never heard of it" but they'll write you a fat ticket if they catch you there.

Also at an undisclosed location in Death Valley is a small puddle against the side of a rock, apparently. Look and you'll see a species of desert pupfish found nowhere else on Earth. Stick your arm in and you'll feel no bottom. Slide your whole body in and you'll find that it's the tiny opening to a vast underground water system from which at least three divers have never returned.

Explorers have also died inside the old Gunsight Mine, as they have in many such mines from California's gold rush and silver rush days, so publishing their exact locations is discouraged. In many cases the bodies are not safely recoverable, as a trip inside will make forbiddingly clear. Whether you learn the old legends about these lost mines or not, their exploration is a trip you'll never forget. If you're lucky you might even photograph orbs inside, like I did, which you can see on the Skeptoid.com web site.

Then there are the natural oddities in California that anyone can visit. The San Joaquin Valley, running down the length of the state, is now plowed into farmlands but it used to be covered with strange rocky lumps called hogwallows. They can be up to 6 feet high and 20 feet across but are often much smaller, though they're always steep and densely packed. Here and there you'll still see an unused corner or a small valley covered with hog wallows. They're so dramatic that they look like they have to be man made, but just about every possible explanation for their existence has been ruled out.

Grimes Canyon in Ventura County is one of only a few places on Earth where rock has been melted by non-volcanic natural processes. It's called combustion metamorphism, and in this case the heat source was decaying organic matter underground. Temperatures reached 3000°F, enough to melt the rock and create a black obsidian-like glass. You can drive right up and a short hike will take you right to this rarest of geological oddities.

Now some of these places I've already been to, as you know if you've watched the videos on the Skeptoid.com web site. I love the unearthly weirdness of the Trona Pinnacles, a huge garden of rocky tufa spires up to 140 feet tall, which grew when mineral-rich groundwater surged from vents hundreds of centuries ago when the region was under water. It's like finding Bryce Canyon in the middle of a vast dry lake. Bring your energy crystals and you can join a healing ceremony there.

I love the Anza-Borrego desert, home to everything from ghost lights to marauding skeletons to lost Viking ships that sailed up the Gulf of California and somehow ended up buried in the desert canyon walls (according to not-very-plausible legend). My favorite place to go is the Mud Caves, a network of dry underground riverbeds that twist for miles beneath the dusty badlands, only of handful of which have been discovered. Said to be a hotbed of floating ghost lights, they make for unforgettable exploration regardless of whether you actually encounter a spook.

Death Valley's Racetrack Playa has always been near and dear to me. In the first volume of this book was a whole chapter devoted to this vast mud flat, crisscrossed with the tracks of rocks that slide mysteriously across the surface all by themselves. How does it happen? Two friends and I witnessed the responsible force, and we found out.

California has an enormous wealth of destinations that, unfortunately, are more interesting to read about than to visit. For example, I visited the Whaley House in San Diego, an old building with a colorful history that's officially registered with the National Park Service as America's Most Haunted House. But when you go there, big surprise, you don't luck into an encounter with the ghost of Yankee Jim, or any of his cohorts. You come home with a disappointingly uninteresting videotape. The fact is that a lot of these cool places have attractions that you're not actually going to see. For many years I planned a Bigfoot hunting expedition. I even sourced a place to rent a thermographic video camera. I plotted out the location of a lake right in the center of the hotspot for many historical sightings. I shopped around for backpackable kayaks and planned to anchor in the center of the lake, and spy all night on a pile of freshly cut fruit on the shore. What stopped me from doing it? The simple fact that I wasn't actually going to see a Bigfoot, or a Loch Ness monster, or an alien. Sure it would have been fun, with the right company and the right flask of recreational beverage, but it's a lot of money and a lot of hassle for what would amount to no more than an extended Happy Hour.

But there may be another creature out there in the murky coastal redwoods of northern California: a giant salamander, the size of a man, lurking in the depths of rivers and streams. It's said to have been spotted many times in the past hundred years but nobody's ever managed to photograph or capture one. If it sounds hard to believe, consider there is a real such animal, at least there is across the pond. The Japanese giant salamander (Andrias japonicus) grows up to six feet long, and it's about the ugliest damn thing you've ever seen. If it can live there, why shouldn't it be able to live in the nearly identical climate and geography in the Pacific Northwest? The only reason I haven't gone on my own giant salamander hunt is that I know full well I would spend three days scrambling through dense poison oak, slipping on river rocks, and freezing to death at night, and never see salamander hide nor hair. The projected return on investment does not reach the tipping point.

There's another opportunity I've always managed to put off. I used to live right next to the famous Claremont Hotel in Berkeley, which boasts a haunted room. I called the reservation desk and asked to rent room 419, but was told you can't request a specific room until after you already show up and pay. They're often full and it's unlikely that a given room will ever be available on a random night. So that was a bust. But the fact is that even if you do get into the haunted room, nothing's going to happen. I was hoping to make a video about my stay there, similar to the other videos on the Skeptoid.com web site, but I wouldn't have ended up with anything to show for it other than maybe an interview with the bartender telling his version of what the laundry woman told him. Yippee.

Then there are the places I don't want to go because they're dumb. Everyone always tells me "Oh, you're in California, you have to go the Mystery Spot in Santa Cruz." The Mystery Spot is an attraction where they've built a wooden cabin at a steep angle, so that balls appear to roll uphill and people can stand on the walls. They've invented a cute story about a gravitational anomaly that has "baffled researchers". If you're six years old it's probably pretty neat; if you have half a brain, "insulting" is a better word for it.

But one of the greatest and most famous haunted locations is always a blast to explore, ghosts or not. It's the Queen Mary, moored in Long Beach. They have an expensive night time ghost tour with a psychic, but if you know your way around, you'll find much of the ship unlocked, fully accessible, and unguarded. A friend of mine was unofficially toured around below decks by another friend who knows the whole ship inside and out. When they reached the haunted first class swimming pool deep within the bowels of the ship, they both felt the "rush of a spirit" burst past them and they ran out terrified. That one's definitely on my list: Even if you don't have a strange experience, it's pretty darn cool to climb around the bowels of a deserted classic old cruise ship in the middle of the night.

Likewise with Hearst Castle in San Simeon and the gigantic Winchester Mystery House in San Jose. Both are renowned for their ghosts, but both are also fascinating in the light of day. Both offer nighttime ghost tours as well as normal daylight tours. Both have impressive histories that feel almost palpable when you're standing right there. The best way to see them is to get to know one of the tour guides who lives onsite, and solicit an invitation to come by after hours. Offer to bring margarita ingredients and you're in like Flint.

If you're a pilot or know a friend with a small plane you can fly over the Blythe Intaglios, a collection a huge figures drawn on the desert floor that are identifiable only from the air, and are believed to be as much as 2000 years old. The Quechan Indians made them by turning over the rocks on the surface, hiding the side blackened by desert varnish and exposing the natural rock colored underside. Desert varnish takes so long to form on a surface that the Blythe Intaglios will be clear as day for thousands of years yet to come. Sadly they've been vandalized pretty badly, and proponents hope to eliminate all ground access to the area. If you want to consider their skeptical angle, you can ponder the claim that the largest figure is pointing directly toward Area 51.

But what a fascinating state we Californians live in. Of course I haven't even remotely scratched the surface, so maybe this is a subject that we'll revisit in the future.

6. Email Myths

You'll never believe what I got in my email this morning. Apparently Microsoft is going to send me $1 for every person I forward this email to! I didn't believe it at first, but then I learned that USA Today featured two whole pages about it. I'll get to it soon — but first little Jimmy with cancer is trying to set a world record for emailed Get Well wishes before he dies, so I've got to take care of him first. I'll just email my encouraging message to the address he provides, addmetospamlist@russianporno.com. I'll do it as soon as I've finished reading Robin Williams' poignant essay on patriotism.

Now, of course, the bastard granddaddy of email myths is the con known as the Nigerian 419 fraud, named after the section in the Nigerian criminal codes that it violates, which is superfluous because Nigeria has never prosecuted a single person for it. Nigerian 419 fraud has been Nigeria's second largest industry, after oil, since the 1980's. This is where you get a politely addressed email from a stranger, often a spouse, family member, or personal attorney of some African leader who has died, full of apologies for bothering you, stating that you seem like a trustworthy person, and for some generous piece of the action you can help them transfer some outrageously huge 8-figure sum out of the country. Of course it's just a con to get your advance fee payments. You may remember the woman who murdered her minister husband not too long ago, in part because she was afraid he'd find out how much money she had lost to this scam.

But most fraudulent emails are harmless hoaxes, or in many cases, accidental mistaken attributions. One of the favorite victims of these emailed essays is comedian George Carlin. If you use email at all, you've almost certainly received one or more essays attributed to George Carlin. One is called Hurricane Rules, a rant against the people of New Orleans, that's halfway decently funny. There's another one variously titled the Bad American or the Bad Republican which is completely random and pointless. There's a transcript of Bill Maher's New Rules from various episodes of his TV show which somehow got compiled and attributed as George Carlin's New Rules for 2006. There's a sappy social critique called Paradox of Our Time. There's a piece called The Stupid Sign suggesting that stupid people should be required to wear a sign identifying them as such. But these are just a sampling; the list goes on and on. In short, virtually every random essay commenting on some aspect of American society is eventually attributed to the late great George Carlin, and this doesn't piss off anyone as much as it did George himself. George finally had to take matters into his own hands and put up a big disclaimer on his web site:

> Because most of this stuff is really lame, it's embarrassing to see my name on it.
> And that's the problem. I want people to know that I take care with my writing, and try to keep my standards high. But most of this "humor" on the Internet is just plain stupid. I guess hard-core fans who follow my stuff closely would be able to spot the fake stuff, because the tone of voice is so different. But a casual fan has no way of knowing, and it bothers me that some people might believe I'd actually be capable of writing some of this stuff.
> Here's a rule of thumb, folks: Nothing you see on the Internet is mine unless it came from one of my albums, books, HBO shows, or appeared on my website.

There's a hardline conservative email essay called Robin Williams' Peace Plan, in which he advises getting rid of the UN, shipping all our illegal aliens to France, deporting foreign students, increasing domestic oil production, and ending foreign aid in a 10-point plan. If Robin Williams, who lives in San Francisco, were to write a peace plan, it would probably look nothing like this one, and it would probably be funny as hell. What seems to have happened in this case is that someone appended a real joke from Robin Williams as point #11, with his attribution, and from then on as the piece was forwarded around it looked like his name at the bottom was for the whole thing. This same mistaken attribution has happened to other people as well. Don't believe attributions on emails.

There is another popular hoax email that promises some reward, like a gift card or even cash, if you simply forward the email to a large number of people to assist with some experiment or test of some email tracking system. No such system exists. Today there is no practical way for anyone to determine whether you forward an email, whether anyone views it, or how many times "around the world" it has gone. Even if Bill Gates wanted to give everyone $1 for each email they forward, current technology prevents him from having any way to know who to pay or how much.

Hoax emails will often say something like "This was featured on *60 Minutes* this week" or "This had 2 whole pages in *USA Today* devoted to it." These are simply lies. I know you've seen that in emails you've received, and I know you've never gone to those sources to verify the claim. The people who send these know that nobody's going to verify anything so they make whatever ridiculous claims they want. "This email has gone around the world 3 times" — what does that even mean, and was that appended to the letter before it was sent or by some mid-level recipient? How was it determined? And who added all those testimonials, and when were they added? Did you add one? Did the person who sent it to you add one? Or is it possible that whoever started it wrote it all and is still laughing?

With all those emails flying around the world out there, the government better get in on the action and grab a piece of it. There have been perennial emails for 20 years now warning that the U.S. Postal Service is going to charge a 5¢ fee on every email, or that Congress is trying to pass bill 602P to tax email delivery. This hoax has happened in the US, Canada, Australia, and probably other countries too. There's no truth to any of it, and the names, law firms, and bill numbers in the original emails were completely fictitious. No US government body has ever pondered such a thing. But the rumors fly so much that even Hillary Clinton was famously fooled by it, and spoke out against the non-existent bill on live television during the 2000 Senate race.

So back to the sick little child who hopes that you'll forward their email to set a record. Other versions of this groundless hoax claim that the American Cancer Society will donate a dime for every person you forward the email to. "All forwarded emails are tracked to obtain the total count." As we know it's not possible for all forwarded emails to be tracked. There are many, many variations on this theme where some company or celebrity promises to donate money for some sick child based on the number of emails forwarded, when in fact there is no possible way for such a count to be derived. You've probably seen a version of this email that starts with a touching little poem called *Slow Dance* written by a terminally ill little girl, advising us to slow down and smell the roses. That was the theme of the poem, but in fact it wasn't written by a little girl, it was written by a perfectly healthy psychologist who had dealt with such children. The name of the little girl in the email is a fictitious one. It's a touching poem misused in a stupid hoax.

As long as we're talking about stupid or harmless hoaxes, be aware that there are even stupider harmful hoaxes. One that's particularly messed up is a legitimate email about a real missing child, including her picture, and all the real information about the case. Sounds like a great use of email, right? Well, it would be. But in some cases, people change the contact information or phone numbers so that in the event the email is received by someone who may actually have helpful information to pass along, they end up calling some premium rate telephone number that has nothing to do with the missing child hotline, and end up surprised by a fat charge on their telephone bill. The lesson to be learned from this is to always be skeptical of contact information in emails asking for help. If you actually have some helpful information, call your local police department to be sure your information goes to the right place.

And of course there are stupid hoax emails promising extraordinary events, like Mars appearing as large as the full moon in the sky. This one's been going around forever and still makes an appearance every year. "It will look like the Earth has two moons," the email promises. The original basis for this was an event in 2003 where the Earth and Mars actually did make a very close pass, and Mars actually did look as large as the full moon — so long as you were looking through a 75× telescope. The current iteration of the email typically leaves out that little detail. Remember: When you read a promise that sounds too good to be true, or that seems to violate the laws of the universe, you have very good reason to be skeptical.

How about that other email that just won't die: the warning that your cell phone number is about to be released to telemarketers unless you sign up for a cell phone Do Not Call registry. The Federal Trade Commission has talked themselves blue in the face trying to advise everyone that this is a hoax and it's simply untrue. Telemarketers are barred from calling your cell phone without your consent. There is only one Do Not Call registry, it's found at DoNotCall.gov, and although you are welcome to put your cell phone number into it, there is generally no reason to do so.

I'll leave you with one final email myth, one you may have heard about recently. It's a proposal for consumers to take charge and actually reverse our high gasoline prices. This chain email, which purports to have originated from an economist or Fortune 500 financial executive, suggests that everyone boycott gas stations owned by the two largest oil companies, Exxon and Mobil (actually Exxon and Mobil are the same company now, the two biggest, at the moment, are really ExxonMobil and Royal Dutch Shell). This boycott will force the largest companies to drop their prices more and more, finally forcing the little guys to drop their own prices to compete. In this manner, consumers can actually wield final control over industry prices. Sounds great, doesn't it? Everyone likes to hear that the big corporations can be bullied around by a guy on the street. Well, you're welcome to give this a shot, if you think it will do anything. Oil companies don't own the gas stations. They're franchises, owned by your neighbor down the street, your local small businessperson. Boycotting his station is only going to put your neighborhood buddy out of business. The competing gas stations, who are now getting all the business, will have to meet the new demand and will simply buy their gasoline from the boycotted producer's refinery. Oil companies do this all the time. It's a commodity, and it's sold on the commodities market, and they constantly buy from one another to meet changing demand. This boycott strategy, according to at least one real economist who has analyzed it, will have no effect on the big oil companies. Its effect will only be to injure the small independent station owner, and to actually raise prices at the busy gas stations who are struggling to meet the demand. Remember, simple chain emails are very unlikely to be the solution to problems that economists and professional analysts have been wrestling with for decades. When you see an unrealistic promise that seems too good to be true, be skeptical.

The take-home from all of this is that you should read every chain email with a grain of salt. Very few of them are true. Virtually all of them making heartwarming promises are false. There are plenty of perfectly good jokes and videos for people to send around that are well worth all of my spare email reading time, that I don't need any of the promises from Bill Gates or warnings that the solar system is about to be dissolved in a nebula. Email petitions are worthless. Emailed promises or claims are rarely worth the electrons they're printed on. Help cleanse the Internet of useless noise. Don't forward chain emails.

7. Fluoridation: Death from the Faucet!

Today we're going to wrap our big juicy lips around the kitchen faucet, turn on the valve, and fill our bodies with a poisonous chemical placed in our water by the government: fluoride.

Most people understand that fluoridation of water means that fluoride is added by the local municipal water supplier, and that's generally correct. What most people don't know is that in some cases, fluoridation means *removing* excess fluoride that occurs naturally in the water supply. Fluoride is a natural component of groundwater, and it occurs naturally everywhere in the world, in varying amounts. The process of fluoridation is to adjust the fluoride content of the water to the most healthful level.

So how did fluoridation become a normal part of municipal water supply? It all goes back to an early 20th century dentist named Dr. Frederick McKay, who practiced dentistry in Colorado, and noticed that a lot of his patients seemed to have brown teeth. In Texas, brown teeth were so prevalent that they were simply called "Texas Teeth". Dr. McKay spent 30 years investigating the cause. Why? Because it also turned out that people with Texas Teeth also had extremely low levels of dental decay. If you had brown teeth, you were only 1/3 as likely to have cavities.

Finally, in 1931, it was determined that naturally occurring fluoride in the local drinking water was responsible for both the discoloration and the lack of decay. Texas and Colorado had extremely high levels of natural fluoride, causing the discoloration, a condition now known as dental fluorosis, which is harmless if a tad unattractive. Years of research and testing in different cities and states, conducted by the National Health Service, determined that one part per million was the ideal proportion, giving the same protection from decay, and avoiding the dental fluorosis. Ever since then, it has been the standard practice to regulate fluoride levels in municipal water supplies to one part per million. There has been broad scientific and medical consensus for decades that one part per million of fluoride is best for health, and exactly zero rigorously conducted scientific trials that have indicated any sign of danger. For all practical purposes, it is an over-and-done-with issue.

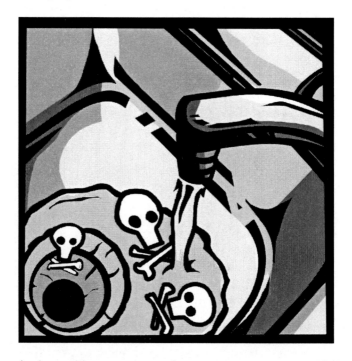

And yet, like so many advances in science or medicine, fluoridation is criticized by a small yet vocal fringe group. There is absolutely an anti-fluoridation lobby in this country. Their process is to flood the mass media with as many claims as they can invent: Claims like fluoridation causes cancer or other illnesses; that insufficient research has been done or that there is "scientific controversy" surrounding fluoridation; that fluoride is a dangerous chemical poison; that fluoridation has been banned in Europe; or any of a dozen other baseless and untrue statements intended to alarm and frighten the public. Alarming the public is not hard to do. There are many communities in the United States where voters have been compelled to ban fluoridation by this widespread misinformation campaign.

Let's turn our eye onto one such community, Arcata, an idyllic coastal hamlet in northern California, that recently won this battle after a divisive and painful fight in the newspapers and in city hall. A principal champion of the science behind fluoridation is Kevin Hoover, editor of the Arcata Eye newspaper. In answering the flood of anti-fluoridation scare tactics, Hoover said:

> *There are no known victims. If there was a problem with municipal fluoridation, wouldn't we have at least a few people who showed some signs of harm after 44 years? All the anti-fluoride people could say was that the victims are "undiagnosed," but not why. They produced no victims, just lots of dubious statistics and horror stories with no provenance.*

Measure W to ban fluoridation was carefully crafted by the anti-fluoridation lobby to simply require FDA approval of anything added to Arcata's water supply, which sounds reasonable and sounds like a good idea, and a layperson otherwise uninformed would be likely to vote for it. The catch is that the Food & Drug Administration has nothing whatsoever to do with municipal water supplies, and so of course FDA approval would never happen, by law. Measure W was essentially a devious, deceitful trick intended to further the anti-fluoridation lobby's agenda at the expense of the dental health of Arcata's children. Generally, it's this same tactic that has been responsible for most anti-fluoridation measures that have passed in the United States.

How else does the anti-fluoridation lobby go about spreading their misinformation? Generally they distribute an eight page pamphlet written by Dr. John Yiamouyiannis, the grandfather of anti-fluoridation activism. Dr. Yiamouyiannis was a naturopath who rejected modern medicine, and was the principal originator of the claim that fluoridation causes cancer.

He raised his family with an emphasis on a fluoride-free diet to avoid cancer. And, as I'm sure you've guessed, Dr. Yiamouyiannis died of cancer in 2000, which he had refused to treat in accordance with his naturopathic philosophy. His type of cancer has a 95% 5-year survival rate, when properly treated.

Most other experts cited by activists are people like Dr. Hugo Theorell, who did indeed oppose fluoridation in the early days. What they don't tell you is that Dr. Theorell changed his mind and became a supporter after the research was published. They'll often cite Swedish Nobel Prize winner Arvid Carlsson, known for his work with dopamine. He's the only known Nobel Prize winner to oppose fluoridation, but the activists multiply him and frequently say that "dozens" or "many" Nobel Prize winners oppose it. When you can only find one guy who opposes something, and his work is in a completely different field anyway, that's a pretty sad commentary on your position. It's also a case of the exception proving the rule. There are always a few contrarian scientists in every field with opinions opposite from the consensus.

It's also stated that fluoridation adds dangerous levels of lead, arsenic, and mercury to the water. Again, this is simply untrue, and making such a claim is really a form of terrorism. In Arcata, no detectable levels of any of those are found in the fluoridated water. Not just below safe levels, mind you; zero.

You'll also hear the claim that fluoridation has been banned in Europe. This is also completely untrue. In Europe it's more common to fluoridate salt instead of water, thus bringing the same benefits via a different delivery method. As long as you don't look at that fact, the anti-fluoridation people can truthfully say that "Europe rejects fluoridation of water."

Thanks to the efforts of Hoover and all of Arcata's doctors, dentists, educators, social workers and newspapers, Measure W to ban fluoridation was soundly defeated in the election. And it's a good thing, too: according to sources in Arcata, if Measure W had passed, the same people were going to try and ban childhood vaccinations next.

Why do they do it? We can really only speculate. Presumably most of these people are good citizens who love their families and want the best for everyone. I speculate that a lot of them are simply ignorant of the facts, and possibly mistrust of the government or anticorporatism compels them to tend to ignore information from official sources and embrace alternative claims, whatever their source. Hoover gave his own answer to this question in an editorial for the Arcata Eye:

> *Billion-dollar industries thrive around entirely imaginary "phenomena."* Astrology, numerology, UFOs, alien abductions, Holocaust denial, the face on Mars, "chemtrails," innumerable media-centered conspiracy theories and fluoride-phobia thrive because they inhabit that magical nexus where paranoia meets superstition – fertile ground for fomenting fear.

The United States Public Health Service estimates that every dollar spent fluoridating water saves fifty dollars in dental expenses. If fluoridation is truly just another conspiracy, then at least this is one that saves money.

8. Who Are the Raëlians, and Why Are They Naked?

A journalist once shouted "The Raëlians are great material: They're sexy, good-looking nudists, and they worship space aliens!" If that doesn't get your attention, you're probably dead.

So who are the Raëlians? Most people have heard of them but few know much about them beyond some vague reference to space aliens or cloning. The Raëlians were founded in 1973, by a young French street musician, race car driver, and automative journalist named Claude Vorilhon. In December of that year, he was hiking around in the crater of an extinct volcano in France and claims to have been greeted by the sight of a flying saucer coming down and landing. A little alien came out and approached him. The alien looked like a short Japanese guy and fortuitously spoke French (French being the universal language, as we know). For an hour each day over six consective days, the alien, whose name was Yahweh, met with Vorilhon and explained the true history of the Earth.

Evidently, Yahweh's alien race were called the Elohim, which also happens to be a Hebrew word with a variety of vague meanings pertaining to holiness or divine beings. Raëlians interpret the word as referring to the alien creators of humanity who came from the sky. Over the course of the six days, Yahweh explained all of the events of the Old Testament as being actual events with purely natural explanations, usually assisted by the Elohim. Adam and Eve, for example, were a literal man and woman who came from DNA custom designed by the Elohim. Humanity is literally the implanted descendants of the Elohim.

And, like the Christian second coming, Yahweh promised Vorilhon that the Elohim would return to Earth when enough of its people have learned the true history of their race and have become peaceful and prosperous. For this purpose, Vorilhon was instructed to build an embassy at which the Elohim can be received when they show up.

Well, as you can imagine, Vorilhon must have been pretty well floored by this startling experience, so much so that he adopted the name Raël, with an umlaut over the *e* to make it look more exotic, and set forth to found the organization specified by Yahweh. In time this became the International Raëlian Movement.

Two years later, Yahweh kicked it up a notch. He came back, and this time picked up Raël and took him for a ride in his flying saucer. They first stopped at an orbiting spa in the outer reaches of our solar system, where Raël received a relaxing massage and aromatherapy treatment — and I swear to you I'm not making this up. This is the doctrine they believe. They then proceeded to the Elohim's home world, a warm jungle planet called the Planet of the Eternals. Raël was introduced to Jesus, Buddha, Mohammad, and Moses, and the group went out for a nice dinner. Raël was then shown their facilities for creating immortal biological robots. They created one of his mother, and a whole group of attractive young females, and they all went back to Raël's apartment and had a party. Raël has written extensively about his climax at the party. Please read these accounts yourself if you think I'm in any way exaggerating or coloring this. All of Raël's books are freely available on the Internet.

The next morning Raël was taken to another facility where they used a helmet machine like something out of a Star Trek episode to "maximize his faculties", presumably making him as intelligent as possible. Yahweh then explained the importance of this procedure: It was to best prepare Raël for geniocracy, a form of government where intelligence determines who should lead. Yahweh then flew Raël back to Earth. The total time between Raël's first encounter and his return from the trip into space: 666 days.

By this time, there were a few hundred members of the Raëlian movement. Over the last 30 years or so, this number has grown to somewhere in the tens of thousands, though estimates vary. Japan and Korea in particular have taken quite kindly to the Raëlian movement, and you'll find some of their largest chapters over there.

So now let's look at all of these tens of thousands of people a little more closely, to find out who they are and what they believe. Significantly, they embrace technological advances to improve the human condition. They support genetically modified crops to best feed the growing population. They love advanced materials science and nanotechnology. They support nuclear power and fusion research for cheap, clean energy. They support genetic research and manipulation to produce people who are as healthy and long-lived as genetically possible. They support terraforming of other planets as the technology becomes available. Much has been made of the company Clonaid, which is owned and operated by Raëliens, and claims to have already created at least one human clone, amid great scientific and ethical controversy. The child is said to be a girl named Eve living in Israel. Clonaid's mission is to create human clones that grow rapidly as spare parts and new bodies for aging humans. Transfer your intelligence into your new cloned body, according to procedures detailed by Yahweh that involve advanced computer backups, and suddenly you're 25 again.

Philosophically, Raëlians oppose war and violence in all its forms. Raëlians frequently appear at anti-war rallies dressed as aliens (which seems incongruous since they say that aliens look like humans). Raëlians also embrace free love and nudity. They consider love to be the answer to virtually all of the world's problems. It's been reported that at some Raëlian conferences, attendees wear colored wristbands indicating what gender or genders they'd like to have sex with tonight. Nudity is the de facto uniform at many Raëlian retreats. I have heard from more than one source that sexual tourists to Asia are catching onto the Raëlian movement. Show up in Japan or Korea, find a local Raëlian chapter and pretend to be a fellow Raëlian, and you get all the sex you want; plus it's free and cleaner and safer than what you get from the back alley brothels. There are affiliated Raëlian organizations dedicated to free love: Raël's Girls, for example, is comprised exclusively of atheist female adult film stars dedicated to Raël. Their web site says "Raël's Girls want to share the understanding that you can be spiritual, you can be happy, without the guilt of God and religion."

Take the perverted poseurs out of the equation and just focus on the true Raëlian philosophy, and many atheists and skeptics will find a lot that they can get on board with. Science as a solution to the problems of the world is a great strategy. Advanced technology to improve health and longevity is a great thing too. Love instead of war, who can argue with that? Moreover, the Raëlian Movement is probably less harmful than many other religions, since they don't teach that your child can be cured of his cancer by telepathically appealing to a paranormal superbeing, and so far nobody's ever fought a war or perpetrated a terrorist attack over Raëlian religious differences.

So if it's that great, why don't we all rush right out and become Raëlians today? The answer to this is similar to what I gave when we discussed neopagan religions on the Skeptoid podcast. There might be a lot to recommend about many aspects of Raëlian philosophy, but the problem comes when you examine the underlying dogma. The statement from Raël's Girls actually sums it up quite well: that you can be happy and support science and technology and love, without the guilt of God and religion; though I'd take it a step further and say that you can have all that without the fiction of Raël's ridiculous alternative creation myth and stupid alien stories. You can have all that without adopting an untrue belief system: The Elohim, Raël's spaceship rides, and the $20 million alien embassy outside Jerusalem. The Raëlian story of creation cannot be reconciled with what we know of evolutionary biology and our planet's geological development. Fundamentally, true Raëlians do what they do and believe what they believe because they see Raël as a prophet who brought back an alternative creation myth from outer space. And by doing so, they are setting aside reason and rationality. And when they go forward with their cloning research and their genetically modified crops and their free love retreats, they are doing it without reason or rationality. And that's just plain dangerous and dumb.

My guess is that a lot of Raëlians are young people who are attracted to the free love. And that's fine, but when they grow up and found companies like Clonaid doing real experiments on real humans, it's time to grow up for real and formally reject the irrationality of Raëlian dogma. The Clonaid people haven't done that. Even if you support genetic research, and a lot of skeptics do, companies like Clonaid who do it for the wrong reasons should be regarded as dangerous, dangerous animals.

9. Will Drinking from Plastic Bottles Kill You?

Today we're going to place our plastic water bottle, which has already been used three or four times, in the car on a hot sunny day, and then drink its noxious chemical contents to see if we get sick and die. The idea is that chemicals in the plastic get released into the bottle's contents when the bottle is reused, and especially if it's heated up.

So let's point our skeptical eye at the issue and see whether it has any merit. Do we need to be concerned about this? The only really fair answer is that it's a complicated question. "Plastic" is not a single compound. There are almost as many different types of plastic as there are types of substances contained by them. Some plastics do contain poisonous chemicals. Some plastics do leech chemicals into liquids. In some plastics, this process can be accelerated by heat. The reason for this variety is to provide the product distributor with enough choices that they can select a plastic type that's best for their product. This permits a distributor of drinking water to use a bottle that is absolutely safe to contain water for humans under the whole temperature range that the bottle is likely to be subjected to. But, put gasoline into that same bottle, and you might see that plastic dissolve away. Plastics are designed for their particular application, and misusing a plastic product can produce undesired consequences.

One time, in college, I was moving to a new apartment a block or two away. My brother and I had built a koi pond, and we needed to move the fish and store them long enough to build a new pond at the new place. We went out and bought a cheap plastic kids' wading pool. We put it in the garage and filled it with the hose, treated the water with all the usual fish-friendly chemicals, and walked the koi over in buckets and placed them in their new temporary home. Well, we learned a harsh lesson about chemicals in plastics. After a day or two the koi didn't look so good. Some of them died. Then all of them died. It was pretty horrible, because, and I'll spare you the details, they didn't look very good. We had no idea what the problem was. Was it the shock of being transported? Did we not add enough stuff to kill the chlorine? On a whim I called the manufacturer of the swimming pool and asked if they knew any reason why this would happen. They did. On products like this, they always add a mold inhibitor to the plastic. In this case, they used cyanide. For a children's pool, they add a safe low level of cyanide that's harmless to the children, but is enough to prevent mold from growing that would make the pool gross and unsightly. Evidently, a level of cyanide that's safe for a human is lethal for a fish, since they breathe it directly into their blood through their gills. The guy we spoke to was the company's head scientist, and he seemed to relish this rare opportunity to discuss his work. He went into all sorts of detail about their different products, and how they use the right plastic for each different job. Ever since then, whenever I work on a koi pond, I always call the manufacturer of any plastic products I'm using and talk to their chemists.

Here's the long and the short of it. Whether you're microwaving food in a plastic container, refilling your plastic water bottle, or making a koi pond, use plastic products that are intended for that use. The manufacturers do employ chemists to determine how best to package their products to ensure their safety, this process is strictly policed by the FDA, and this is always going to be more reliable than random information you read on the Internet or receive in a chain email.

And yes, it is our good old friend the Internet that seems to be the basis for this particular fear's place in popular culture. For example, there's one hoax email going around that says Sheryl Crow believes she contracted breast cancer from toxic chemicals by drinking water from a bottle that had been left in a car. Not true. Sheryl Crow doesn't claim this, there are no chemicals in water bottles that have been linked to cancer, and heating a water bottle to car temperatures does not leech anything into the water. There's another chain email that says freezing your water bottle, like so many people do, will leech dioxin into your water. Again, not true. No plastic containers designed for containing food or drinks contain dioxin, and colder temperatures stabilize plastics; it's heat that will accelerate their breakdown.

Most famously, a 2001 study at the University of Idaho found that reuse of plastic water bottles does release risky levels of diethylhexyl adipate (DEHA) into the water, which is potentially carcinogenic. This study was widely reported by the popular media and largely touched off the chain emails and most of the current perceived controversy. But is it true? No. Such a paper was written, but it was not a formal study. It was, in fact, merely the master's thesis of one student. It was not subjected to any peer review, and cannot accurately be characterized as a study performed by the university. It does not represent any position held by the University of Idaho. And unfortunately, it was not well performed research. DEHA is not classified by the FDA as a carcinogen, but more importantly, DEHA is not used in the type of plastic water bottles that the student evaluated. But it is used in many other plastics, and is present in a lab setting. "For this reason", concluded the International Bottled Water Association (which is, granted, not a very objective source), "the student's detection is likely to have been the result of inadvertent lab contamination." The FDA requires a higher level of scrutiny than that applied by the student writing his paper. DEHA is actually approved for food contact applications, but the fact that it's not present in the type of plastic that was studied, discredits the entire paper. But the mass media is often more interested in headlines than facts, so the dangers of reusing water bottles had no trouble becoming a fixture in pop culture.

Some people allege a conspiracy among distributors of bottled water, who know that their products are poisonous but who have analyzed the cost savings against the projected lawsuits from wrongful death and have concluded that it's more profitable to sell dangerous products. I do not find this theory very compelling. First, the products demonstrably do not contain the toxic agents claimed by the theory. Second, like all conspiracy theories, it's just too implausible that something of that magnitude could be kept secret for so long by so many people and so many victims, with nobody ever blowing a whistle or calling a newspaper. If corporate Men in Black were sent out to silence the whistleblowers and families of the victims, this would just multiply the number of reasons for someone to blow the whistle. This conspiracy theory just doesn't hold any water — pun intended.

There are absolutely plastics that are unsafe for containing or heating food. Look what happened to my koi. Or, let's say you sealed some food inside a length of PVC pipe and heated it over a campfire. Is that safe? I don't know, but I wouldn't eat it. Just like everything else in life, use products for their intended purpose, and you will not have any problem. Be assured that intended use of water bottles does include high temperature cycling. You will not get sick from any reasonable use of a water bottle or other food-containing plastic product.

10. Irradiation: Is Your Food Toxic?

Today we're going to critically examine a mysterious green logo on the side of a food container in the supermarket. Many people believe this means the food is radioactive or poisonous. Some trust that it's simply sterilized. Which is true?

My grandmother went to her grave with the firm belief that eating food that had been warmed in a microwave would give you cancer. The similarities between the terms "microwave radiation" and "radiation poisoning" were all she needed to form this opinion. And, unfortunately, hers was no less rigorous than the process most people follow to form opinions about technologies or methods that they don't thoroughly understand. I offer most Americans' opinion of nuclear energy as a perfect example: Chernobyl = danger = nuclear reactors are fundamentally bad. Yet, how many of them can tell you anything about the closed fuel cycle reactors being designed today? Ever since the aftermath of Hiroshima, there has probably never been any word that incites so much fear as *radiation*.

Simply put, irradiation is the process of blasting food with a shot of ionizing radiation, killing any microbes that would contribute to the food going bad sooner than it needs to. Bacteria, viruses, and everything else are all sterilized by the radiation. Ionizing radiation is used because it's high energy, and is extremely dangerous to living tissue. Three types of radiation are used: high intensity x-rays, which are high energy and penetrating; gamma decay, which is really high energy and does great damage, and consists of electromagnetic radiation in the form of photons; and beta decay, which consists of electrons that are too big and slow to do much by themselves, and so they need to be accelerated into a high energy state at nearly the speed of light. All three types have different characteristics in terms of how far they penetrate and how long an exposure is required, so different types are used for irradiating different foods. Whatever the method, the food coming out the other end — be it bread, milk, meat, fruit, or cheese — is absolutely sterile and, if properly sealed, will last longer on your shelf than virtually anything else in the supermarket.

How is irradiation such an effective killer? The high energy of ionizing radiation creates ionization events within the cells of tissue that it strikes. These ionization events result in chemical and even some nuclear reactions among the affected molecules. When this happens in DNA, it causes damage that, if the cell survives, can become cancer. At higher levels, sufficient damage is done to the cell that it's almost always killed outright. Guaranteed no salmonella or *E. coli*.

Irradiation is a cheaper way to bring safer food to the market. It's common in Europe, where refrigeration infrastructure has historically not been so great, but it's rare in the United States. The circular green logo along with the words "Treated with irradiation" are so terrifying to much of the American public that the process has been put virtually out of business. Most Americans would prefer to accept a few *E. coli* deaths than have their food exposed to a sterilization procedure that involves nuclear physics.

At this point, some of you are tuning out and saying "Oh, there he goes again, towing the line of his corporate paymasters, saying whatever Halliburton is paying him to say, and talking down to people who question authority." Questioning authority is wonderful, and more people should do it. Unfortunately, many of those who claim to question authority are really just rejecting authority, for the sake of doing so. To truly question something, you have to listen, learn, and analyze.

During the anthrax scare of 2001, the US Postal Service had a lesson in analyzing. As an emergency measure, they contrived to irradiate the mail, as this would effectively kill any anthrax that anyone might be sending. Well, apparently they turned up the volume too high. They killed the anthrax, all right, but they also nuked the hell out of a lot of mail. Some stationery darkened horribly. Stamps were ruined, to the dismay of collectors. Photographic film was destroyed. Plans to irradiate all mail eroded, and now only mail to certain government offices gets this treatment. Apparently, you can overnuke stuff. Don't tell my paymasters I said this, but irradiation can be done wrong, and can have disastrous results.

Critics have pointed out that new, unexpected chemicals can be formed during irradiation, and this is true. Ionization is a chemical reaction. However, cooking, and most other food preparation techniques, cause the formation of new chemicals in dramatically higher concentrations, and this has never posed a problem. Every test ever done has found irradiated food to be safe.

But that doesn't mean they're always good. Some foods don't tolerate irradiation well. Some foods, most notably romaine lettuce, can smell bad, taste funny, or have their texture affected if they get overnuked. This is not a health concern, it's a culinary problem. The trick is to find the right dosage to kill any organisms and yet not affect the food. Almost all foods tolerate irradiation with no noticeable effects, but for some, the food gets ruined at doses too low to effectively sterilize it, and so these foods are not candidates for irradiation. For any food you see bearing the irradiation logo, you can be sure that it's been tested and its quality has been found to be unaffected by the dosage used. Not that this will satisfy the more paranoid consumers.

You see, a principal misconception held by many opponents of food irradiation is that the process of applying radiation leaves the food radioactive. This is completely false. This would be analogous to turning off the light in a room, and expecting the room to be residually contaminated with light radiation and still glowing. Obviously, this isn't what happens. Once the source of the radiation is removed or deactivated, there is no more radiation. It's like turning off a light. Food that's been irradiated is not radioactive.

Does this mean that you can take a fish fillet and set it outside your nuclear bomb shelter, let it marinate in the radiation, then bring it back inside and enjoy a healthy meal? No. In this case, heavy radioactive elements in the fallout dust would contaminate the fillet. When you brought it back inside, you'd bring in radioactive dust along with it, and you'll all get fried. The irradiated fillet won't hurt you, but the heavy radioactive metals sitting on its surface, still emitting particles, will. Food irradiation, and for that matter microwaving as well, does not place radioactive material onto the food. The food is placed in the radiation field, and then it's removed. Run a Geiger counter over it, and it shows zero. Food that's been irradiated is not radioactive.

Nobody has ever been sickened or harmed in any way by eating properly irradiated food, despite a few untrue claims made by the Sierra Club and others. However, many people die or catch severe bacterial infections from eating food that is not irradiated. Next time we have an *E. coli* breakout, step up to the plate and point out the benefits of irradiation. When lives are on the line, we should use every means at our disposal to properly sanitize our food supply.

And also use every means at our disposal to get Halliburton to finally send me my check.

11. Crop Circle Jerks

Tonight we're going to take our dowsing rods and our tinfoil helmets, stand out in a remote wheat field, and try to feel the psychic energies as a UFO comes down and forms a gigantic complicated geometric pattern by crushing the wheat. It might use a whirling dimensional vortex as its mechanism. It might be ball lightning or some strange effect of the wind. It might be the aliens trying to communicate with us. It might be the Earth herself expressing profundities. Or it might be a couple of clowns with a piece of wood.

We've all heard how in 1991, two old English guys, Doug Bower and Dave Chorley, went public with the confession that they had been making crop circles throughout England since 1976, using ropes and planks and simple surveyor's tricks. They generally did it after pub night on Fridays, and had a rollicking good time. They had been enjoying the resulting media circus immensely, and would gladly have taken their secret to the grave, but for Bower's wife who noticed the mileage on his car and wondered if he was having an affair. So, to protect Bower's marital bliss, the two made a public confession, and even did live demonstrations on TV. The media reported that the crop circle phenomenon had been solved. But, of course, to any intelligent person, Bower and Chorley's confession didn't prove a thing, any more than Ray Wallace's family's confession about him making Bigfoot prints proved that there weren't also a thousand other sources of footprints. Artist John Lundberg, who formed a group called Circlemakers, has been making many of the most complex and beautiful crop circles ever since the public confession, including many for commercial purposes. Even I made an effort in the late 1980's. My friend and I took some old skis and were going to make a crop circle in Irvine, California, but when we got there we discovered the last remaining field had just been plowed for a new subdivision.

One good thing about the crop circle phenomenon is that there are very few people left who believe that they have some cause other than pranksters. But those few people are resolute in their belief. Hoaxing is now so prominent that most of the staunchest crop circle researchers now concede that the vast majority of crop circles are manmade. However, some of them have found an "out" that lets them continue to stroke their paranormal explanations for even the manmade variety: Some researchers now believe that the same paranormal or alien forces that create "real" crop circles are also responsible for controlling the minds and actions of the hoaxers. Thus even the manmade crop circles are equally significant as evidence that an unknown intelligence is behind all crop circles.

One prominent researcher of crop circles is a man named Colin Andrews, who used to call himself CPRI or Circles Phenomenon Research International until he ran out of money a few years ago. His web site at CropCircleInfo.com offers CD-ROMs and Powerpoint presentations about crop circles for sale, but little in the way of testable hypotheses about non-human origins of crop circles. His research methods largely center on dowsing and psychics — which is what I'd do too, since those sources produce claims of such a nature that they cannot be tested or falsified.

In 1993, Andrews contacted Masahiro Kahata, a Japanese software engineer who constructed a simple device for measuring electroencephalogram activity and displaying it colorfully on a Macintosh screen. He calls it the Interactive Brainwave Visual Analyzer. Kahata came to England, and the two of them tromped around taking amateur EEGs of people on the street as a control, and also of dowsers in the act of examining crop circles. Andrews reported:

> *What we found, measuring with a computer real-time in the fields, was that the right brainwave activity of the dowser, at the precise moment those seven rings were measured and reacted to by the dowsing rods, spiked in all the brainwave frequencies — alpha, theta, beta, and delta — at the precise moment the dowsing rods moved.*

Andrews regards this as hard scientific evidence that the dowsers are reacting to a physical manifestation of the crop circle, though he's vague on what that might be. As it turns out, Kahata had also done similar experiments on his own in the 1970's, when he applied an earlier version of his device to magicians and self-described psychics while they were performing spoon bending tricks. He got the same results: higher EEG activity during the spoon bending performances. Was this evidence of an unknown psychic force? Science writer and magic teacher Dorion Sagan, son of Carl Sagan, offered a different conclusion:

> *If there is a tightly correlated increase in mental activity while a psychic is bending spoons, it is probably because he is nervous he is going to get caught.*

Now I'll grant that most dowsers, especially those who invest the time and money to travel to crop circle sites, are not consciously out to fool anyone and thus aren't nervous that any deception will be detected. But since dowsing of any kind has never passed any rigorously controlled test (sorry, but it hasn't), and it's well established that many psychics and other mediums are genuinely but unconsciously using well established cognitive phenomena to guide their divinations (sorry, but they are), honest dowsers are probably genuinely excited every time their dowsing rods move. And genuine excitement is just as good at making an EEG jump as is the state of being nervous.

Most neurologists agree that EEGs are useful to a certain point. You can derive basic information from them, but they are too vague to indicate anything complex like sending telepathic messages. Intense concentration on a pattern, for example, can produce a recognizable signal in some cases. An epileptic seizure throws up a giant spike. But to state that any given spike indicates the presence of a paranormal force and not the excitement or nervousness of the dowser, you need to leave the realm of what neurologists have learned and enter the land of pure speculation. Andrews himself states that the same spikes in EEG activity were observed on one occasion when a military helicopter flew close by. Such a flyby would make me pretty nervous too.

So much for dowsing the crop circles. What about their formation? The people who make them use simple tools and surveying techniques to transfer complex plans into a full-scale wheat field, but what about those said to be formed by paranormal means? How does that happen? Colin Andrews explains further:

> *The eyewitnesses I've interviewed in many countries over the years have all agreed with me on one point: when they claim to have seen circles form, they appear in 10 to 15 seconds.*

In any picture you see of Colin Andrews visiting a crop circle, he's loaded with camera equipment and so is everyone else in the picture. In fact, it's hard to find any picture of crop circle investigators where everyone in the shot is not holding a camera or binoculars or something, finger on the trigger. So my question to Colin Andrews would be, "Did you not ask these crop circle investigators who witnessed the formations why, in every single case, they failed to produce a single photograph or frame of videotape showing this wonderful creation?" If I were Colin Andrews, these investigators are not those whose testimonials I would flaunt to the world. Instead I would tell them they screwed up, and probably even accuse them of trying to hoax me. How can they spend all day and night camped out on the hilltop, finger on the video camera trigger, witness a crop circle forming, and produce only a lengthy list of verbal reports, and no video? Inexcusable for a conscientious researcher. The first thing I would fault Colin Andrews for would be requiring only the lowest of standards for the information he accepts as evidence.

So what about all these numerous eyewitness accounts of crop circles being formed, in seconds, by hovering balls of light? Well, again, I'd point to the evidence issue. These eyewitnesses, or at least those reporting the accounts, always turn out to be crop circle believers. If they'd seen a real event, they probably would have used that camera hanging around their neck. But in every case, they've failed to do so.

Well, almost every case. There is one famous video of white balls of light actually creating an entire crop circle, in seconds. It's called the Oliver Castle video, and you can find it on YouTube. It was made by John Wabe in 1996 or 1997, a partner in a small video production company called First Cut Studio. He took some simple video of the completed crop circle, and ran it through their Quantel Paintbox. In a video subsequently broadcast on the Discovery Channel and on National Geographic, he showed how he rubber-stamped other pieces of the wheatfield background to "erase" the crop circle, and then un-erased it bit by bit underneath some flying white dots that he added. He then added some shake and some artificial generation loss to the video, and presto, a great hoax was done. For years it was considered definitive proof by many crop circle believers. But when he finally went public with how he made it, guess what? Few believed him, and many still believe to this day that the video is genuine, and that it's his confession that is the real hoax. Web pages accuse him of earning huge sums of money — government payoffs for discrediting a genuine video. Even if you read the comments on YouTube — which are, granted, mostly the half-literate and profanity-laced ravings of young people — it's painfully clear that many people cannot be convinced by any evidence that a paranormal phenomenon is not real.

And although some prominent crop circle researchers (Colin Andrews among them) do accept that the video is a fake, many do not. Believer web sites assert that top video analysts have proven that the Oliver Castle video cannot have been faked. My favorite among these top analysts is Jim Dilletoso. Dilletoso is one of the most vocal UFO advocates, and claims to have spent six weeks at an underground base for gray aliens outside Dulce, New Mexico. Judge his credibility for yourself.

It is an interesting world we live in, where you can tell a group of people that you made a crop circle with a rope, even show them how you did it, and they still insist that an unknown paranormal intelligence did it. You can tell them that two plus two equals four and they'll insist that it's five, even after you line up four apples for them. You can make a simple hoax video with them sitting at your elbow watching, and they'll conclude the video's real and you're a paid government stooge. And then they'll put their tinfoil helmet back on.

12. Subliminal Seduction

Put on your 3D glasses and grab a seat in the theater — it's time for a wild ride through the psychological cinematic world of subliminal advertising.

Alleged subliminal advertising is said to take place in two forms. In the first, a marketing message like "Drink Pepsi" is flashed on a screen so briefly that a person cannot consciously perceive it. In the second, sexual imagery is cunningly hidden within artwork to make it more compelling for no consciously discernible reason. Subliminal means below the threshold of conscious perception. So, for any such message to be truly subliminal, it must not be consciously detectable. In fact sexual imagery is all over advertising, but if you're able to perceive it, it's not subliminal and thus not part of this discussion. Draping a bikini model across the hood of a Camaro is not subliminal advertising.

The magnum opus of subliminal advertising is a book written in 1974 by Wilson Bryan Key, *Subliminal Seduction*, inspired by Vance Packard's original 1957 book *The Hidden Persuaders*. Packard's book discussed ways advertisers might appeal to consumers' hopes, fears, and guilt. Key took it to a whole new level, "exposing" advertising methods that he had envisioned or perceived on his own. *Subliminal Seduction* has been highly successful over the decades, spawning at least two sequels (though they contain much of the same material). Key's assertions have inspired whole college curricula dedicated to propagating the claim that the advertising industry systematically influences the public with subliminal advertising. Just read what a few Amazon customers have said about *Subliminal Seduction:*

"Why You Should Read This Book"
I have studied subliminal techniques for 30 years. You have to understand psychology to appreciate that you can be

manipulated by hidden messages in visual images. Why would commercial artists continue to do it if there were no evidence that it works?

"Read It Carefully"
Key has an uncanny insight into a subject that more people should become aware of. This is where you will miss the boat by not taking the information in this book seriously. The manipulation of our minds is more far reaching than even he could have guessed.

"Enlightening Eye Opener"
This comes from a time before digital computers, and should provoke anyone to ask "If they were that good then, what are they doing to us now?!?"

So what does the advertising industry have to say about subliminal advertising? As it happens, I took a series of advertising seminars earlier in my career with a panel of local ad executives. During one Q&A session, a guy stood up and asked about the findings made in *Subliminal Seduction*. As one, the panel collectively groaned and laughed. They said that book was the oldest joke in the advertising industry. The author Key has never worked in advertising and his books exhibit no practical knowledge of the advertising business, other than his own delusional perceptions of what he sees in ice cubes. Moreover, any ad agency that airbrushed naked women into pictures of their clients' products would find themselves fired very quickly. The student who asked the question showed a magazine ad, and pointed out how some curves in a swimming pool mimic the curve of a woman's back. One of the panelists pointed out that first of all, there is no subliminal aspect to the boldly pictured swimming pool; and second of all, please show us an ad in which you do not find hidden sexual messaging. The student could not. He was firmly convinced that hidden sexual imagery is present in all advertisements; even the curve of a letter S seemed to be in an especially suggestive typeface. All the panelists told him he's wrong and that none of their companies had ever done or seen such a thing. The student would not be dissuaded, and probably concluded that the panelists were covering up an industry conspiracy. He did not return to future classes. He was probably killed by Men in Black for discovering the truth.

By the way, here are a couple more Amazon reviews from people who seem to actually know something about advertising:

"Utter Nonsense"
I worked in New York advertising for five years. Nothing like this was ever done. Key has no concept of what the advertising world is really about.

"The Great Urban Legend"
I've been in the advertising business for thirty years. I can't believe this guy has made a living spreading this drivel for so long. People in advertising continue to laugh about Key. He has no understanding about how advertising works, [or] how the people in it do their job.

But to study the issue with honest skepticism, we must dismiss the advertisers' statements as anecdotal and focus only on testable evidence. So let's turn our eye toward whatever research was done that found subliminal advertising to be effective, and see what justified this student's belief.

Right around the time that Packard's original book was published, a market research consultant named James Vicary set up a special projector inside a movie theater in Fort Lee, New Jersey. Over the course of six weeks, he chose certain showings of the film *Picnic* and throughout those showings, he flashed certain marketing messages onto the screen for .003 seconds (well below the perceptual threshhold), and kept doing it every five seconds through the entire movie. In all, 45,699 customers watched Vicary's movies. The messages said "Drink Coca-Cola" and "Hungry? Eat Popcorn". The result? During movies when Vicary flashed his messages, sales of Coca-Cola rose by an average of 18%, and sales of popcorn rose by 58%.

And so there it was. Since that famous experiment, subliminal messages flashed on TV and movie screens have been a firm fixture in popular culture. Hardly a single student who takes a class on psychology or advertising will escape hearing about it and believing wholeheartedly in the effectiveness of subliminal messaging. The media had a heyday with the sensational headlines, and the rest is history.

Harcourt Assessment, which was known at the time as The Psychological Corporation, invited Vicary to repeat his experiment under controlled conditions. He did, but this time no increases in sales were shown at all. Pressed for an explanation, Vicary confessed that he had falsified the results from his original study. Indeed, five years later in a 1962 interview with *Advertising Age*, Vicary revealed that he had never even conducted the Fort Lee experiment at all. He had literally made up the entire thing. But of course, by then, it was too late. The headlines had run their course, and to this day it's a generally accepted fact that flashing brief messages onscreen produces a desired behavior, despite the fact it never happened.

Is there any evidence that subliminal messages or hidden sexual imagery produces higher sales? Evidently, no. At least, I couldn't find any. However I did find one relevant study from 2007, from the University of California Davis. The findings, surprisingly, were that subliminal sexual images had no effect on men, and actually produced lower levels of sexual arousal in women. Neither group went out and bought popcorn or Pepsi. The conclusion suggests "that the subliminal sexual prime causes women to activate sex-related mental contents but to experience the result as somewhat aversive." Not really a great advertising strategy.

Next time you eat a Ritz cracker, examine it carefully. Wilson Key believes that it has the letters S-E-X stamped on its surface, but in such a way that you can't consciously perceive it. Do try to find it, give Key the benefit of the doubt. And then decide for yourself whether it's actually there, or whether this whole urban legend is just another stupid, baseless, sensationalist headline.

13. The Attack of Spring Heeled Jack

Come with us now to 19th century London on a dark misty night, full of spectral villains and unspoken fears. For this was the realm of Spring Heeled Jack, one of the most popular and frightening characters from recent English lore. A composite description of Spring Heeled Jack was a man with devilish facial features, a frightening grin, glowing red eyes, and a terrifying high-pitched laugh. He wore a tight-fitting white oilskin suit and a shiny metal helmet. With his cape and boots he had quite the superhero look about him. He spat blue flames at will, and most extraordinarily, he could jump in a most demonstrative manner, clearing buildings and high walls with ease, and crossing towns in moments by bounding from rooftop to rooftop. Over a period of decades during the 1800's, he made many appearances, always troublesome and usually malicious, tricking and attacking innocent victims and leaving a wake of terror all across England.

Although Jack's exploits are said to have taken place all over England and to have numbered in the dozens at the very least, there are really only six or ten specific incidents to be found in the literature. When you research Spring Heeled Jack, you read the same half dozen accounts over and over again. There are a couple stories of him knocking at peoples' doors, perhaps with a plea for help, and blowing flames in their faces when they answer; there are a couple cases of him attacking and harassing soldiers on guard duty; some molestations of young women; and there is an episode or two of being shot at by villagers with no effect. In every case Jack would escape with his mighty superhuman jumping, bounding over tall buildings, laughing and cackling like a drunken banshee.

Some believers tend to take these old stories at literal face value, and so come up with wild hypotheses that are the only way to fit all the claims of the story. It's been suggested that

Spring Heeled Jack was an extraterrestrial alien, who was from a planet with high gravity and so had an extraordinary jumping ability on Earth. Our thin atmosphere could have made him giddy, thus accounting for his laughter and wild ways. And his species could have been nocturnal, giving him reflective eyes like a cat that would explain his glowing red gaze. What about his fire breathing? Easily explained as "odorous phosphor," illuminated by his alien bioluminescence or ignited by a bioelectric shock strong enough to stun his victims.

Anyone who's heard of Spring Heeled Jack has probably heard the most common nomination of a suspect: Henry Beresford, the Third Marquess of Waterford, known as the "Mad Marquis" for his mischievous and boisterous nature. He was a contemporary of Jumping Jack, and although his principal home was not near London, his continual drunken partying took him all over England and he did live in the area at the right time. The problem with this nomination is that there was never the slightest shred of evidence linking him to Jack, or even really enough to justify any suspicion. It was said that the Marquess had been embarrassed by women and by the police during his career, and this was his way of getting even. Well, let's count the number of people in England during the 1800's who had been embarrassed by women or by the police. Hmmm. The other weak shred linking him was a little boy's report that Spring Heeled Jack had a W embroidered on his shirt when he appeared at the door, and W could stand for Waterford. When you consider the many names and places that W might stand for, or the many other reports that had Jack dressed differently, there appears to be little reason to support such a connection. Nevertheless, put two things next to each other, and people draw connections and spot patterns. It was said that some of Henry's friends were interested in science. Well, so were a lot of people, and so were a lot of people's friends. But in this case, it was opined that these friends could have designed special spring-loaded boots for the Marquess that allowed him to jump over buildings. Logically, these supposed facts are completely worthless. Factually, Henry Beresford died in a riding accident shortly after the first of

Spring Heeled Jack's appearances. In all of my research, I found not a single reason to support the Marquess of Waterford hypothesis. Sure, maybe he was guilty, and maybe my cat was too.

So what does our skeptical eye see when we turn it toward Spring Heeled Jack? Surely there wouldn't be all these long-enduring stories unless they had some basis in fact.

I'd like to turn the clock back for a moment to early 2001. Let's spin the globe and place our finger on New Delhi, India. Picture great masses of humanity moving through the dusty heat. Imagine a busy marketplace, a bustling trade district of glass skyscrapers with smoking motorcycles, pedicabs, wall-to-wall apartment buildings, tangled bunches of telephone wires, and everywhere you look, people, people, and more people. In this melting pot of cultures, languages and economies, a mysterious creature called the Monkey Man came out of nowhere and terrorized the nation's capital for three months. Police received 350 reports — a number that dwarfs Spring Heeled Jack's total — from victims claiming to have been bitten, scratched, and pummeled by a bizarre half-man, half-monkey creature. One hospital reported 35 victims with injuries that appeared to be animal bites. At least two people actually died in falls while fleeing the beast. Police offered a thousand dollar reward for information leading to the capture of the Monkey Man (and a thousand dollars was no small change in India), and even issued renderings made by a sketch artist, that looked a lot like an angry Curious George. Great mobs swarmed into the streets with bricks and bats and anything they could grab to kill the monster, and once they chased a four foot tall wandering Hindu and beat him into a coma before the police could intervene. In another case, a van driver was pulled from his vehicle and savagely beaten. The Monkey Man seemed to be everywhere, jumping out from bushes and attacking the vulnerable. The whole phenomena was uncannily like that of Spring Heeled Jack.

You might ask why, since this happened to a forewarned population in one of the most densely peopled places on Earth over a period of months, nobody ever got a picture or security camera video or any real evidence of the Monkey Man. The injuries treated at hospitals could be called evidence, but the Times of India quoted police sources as saying "In most of the cases, the injuries were found to be too superficial to arrive at any conclusion. Most of the wounds could have been self-inflicted."

The hundreds of eyewitness accounts aren't evidence either, and the police sources explained why quite aptly: "It was found many victims changed their statements on several occasions. Psychiatrists concluded most of them were hysterical and could not be relied on."

When you take a few million superstitious people and flood them with sensational headlines stating that hundreds of people are being attacked everywhere, you can easily get a kind of mass panic, not too different from what the eastern United States experienced during the sniper attacks a few years ago. According to the Hindustan Times, "It was due to unsubstantiated media reports that people were encouraged to come out with bizarre accounts of the creature though no one had actually seen it."

Why were there no pictures? Simple, there was no Monkey Man. We'll never really know what started the craze: Maybe it was a kid with a mask, maybe it was an actual attack or mugging. It may have been nothing more than someone's made-up story, or even a betel nut hallucination. And as for Spring Heeled Jack? A tall tale to explain a ravished young lady? A young lad's explanation for having been beaten up in a pub brawl? A story told from lip to lip until it reached a newspaper reporter? It could have been anything. There is every reason to be skeptical of Spring Heeled Jack having ever existed at all, and neither evidence nor plausible explanations to keep him flying high.

14. How to Argue with a Young Earth Creationist

Young Earth Creationism is a relatively recent movement within Christian fundamentalism, and it's built upon a rejection of virtually all scientific disciplines, since they all converge upon a physical age of the Earth that contradicts a literal Biblical interpretation. Young Earthers accept the Bible as they only reliable scientific document, and point to the age of the Earth as being from six to ten thousands years old. They have invented new versions of virtually all physical sciences to support this claim.

Debating with a Young Earth Creationist is actually really easy, because they only have a few standard arguments, and haven't come up with any new cogent ones for some time. These standard arguments have been published time and time again, and a practiced Young Earther can handily draw them like a six-gun at the drop of a hat. All of their arguments are silly in their wrongness and easily debunked, and if you're prepared in advance, it's easy to beat down any Young Earther with a quick verbal body slam. You're not going to change their mind, since creationists do not base their opinions upon rational study of the evidence; but you might help clear things up for an innocent bystander who overhears.

So here are the standard arguments for Young Earth Creationism, and the standard rebuttals from the scientific consensus, starting with my favorite:

Evolution is just a theory, not a fact.
This is an easily digestible sound bite intended to show that evolution is just an unproven hypothesis, like any other, and thus should not be taught in schools as if it were fact. Actually, evolution is both a theory and a fact. A fact is something we observe in the world, and a theory is our best explanation for it. Stephen Jay Gould famously addressed this argument by

pointing out that the fact of gravity is that things fall, and our theory of gravity began with Isaac Newton and was later replaced by Einstein's improved theory. The current state of our theory to explain gravity does not affect the fact that things fall. Similarly, Darwin's original theory of evolution was highly incomplete and had plenty of errors. Today's theory is still incomplete but it's a thousand times better than it was in Darwin's day. But the state of our explanation does not affect the observed fact that species evolve over time.

Evolution is controversial; scientists disagree on its validity.
Young Earthers have latched onto the fact that evolutionary biologists still have competing theories to explain numerous minor aspects of evolution. Throwing out evolution for this reason would be like dismissing the use of tires on cars because there are competing tread designs. Despite the claim of widespread controversy, no significant number of scientists doubt either the fact of evolution or the validity of the theory as a whole. Young Earthers often publish lists of scientists whom they say reject evolution. These lists are probably true. In the United States, the majority of the general public are creationists of one flavor or another. But the scientific community has a very different opinion: Most surveys of scientists find that 95 to 98 percent accept evolution just as they do other aspects of the natural world.

Evolution is not falsifiable, therefore it's not science.
One of the fundamentals of any science is that it's falsifiable. If a test can be derived that, if it were to fail, falsified a proposition, then that proposition meets a basic test of being a science. Something that cannot be tested and falsified, like the existence of gods, is therefore not a science. Young Earthers accept this to the point that they use it as an argument against evolution's status as a science.

In fact, evolution could be very easily falsified. Evolutionary biologist JBS Haltane famously said that a fossilized rabbit from the Precambrian era would do it. Another way to falsify evolution would be to test any of the innumerable predictions it makes, and see if the observation doesn't match what was predicted. Young Earthers are invited to go through all the predictions made in the evolutionary literature, and if they can genuinely find that not a single one is testable, then they're right.

Evolution is itself a religion.

This argument has become increasingly popular in recent years as Young Earthers have tried to bolster their own position by decorating it with scientific-sounding words like *intelligent design*. And as they try to convince us that their own position is science based, they correspondingly mock evolution by calling it a religion of those who worship Darwin as a prophet and accept its tenets on faith since there is no evidence supporting evolution. Clearly this is an argument that could only be persuasive to people who know little or nothing about the concept of evolution or Darwin's role in its development. This argument is easily dismissed. A religion is the worship of a supernatural divine superbeing, and there is nothing anywhere in the theory of evolution that makes reference to such a being, and not a single living human considers himself a member of any "evolution church."

Evolution cannot be observed.

Part of what you need to do to validate a theory is to test it and observe the results. Although there are evolutionary phenomena that can be directly observed like dog breeding and lab experiments with fruit flies, most of what evolution explains has happened over millions of years and so, quite obviously, nobody was around to observe most of it. This is true, but it misstates what observation consists of. There's a lot of observation in science where we have to use evidence of an event: certain chemical reactions, subatomic particle physics, theoretical physics; all of these disciplines involve experimentation and observation where the actual events can't be witnessed. The theory of evolution was originally developed to explain the evidence that was observed from the fossil record. So in this respect, every significant aspect of evolution has been exhaustively observed and documented, many times over.

There is an absence of transitional fossils.
If the ancestor of the modern horse Miohippus evolved from its predecessor Mesohippus, then surely there must be examples of transitional fossils that would show characteristics of both, or perhaps an intermediate stage. I use the horse example because the fossil record of horses is exceptionally well represented with many finds. If evolution is true, shouldn't there be examples of transitional stages between Miohippus and Mesohippus? The creationists say that there are not. Well, there are, and in abundance. You can tell people that there aren't, but you're either intentionally lying or intentionally refusing to inform yourself on a subject you're claiming to be authoritative on. Kathleen Hunt of the University of Washington writes:

A typical Miohippus was distinctly larger than a typical Mesohippus, with a slightly longer skull. The facial fossa was deeper and more expanded. In addition, the ankle joint had changed subtly. Miohippus also began to show a variable extra crest on its upper cheek teeth. In later horse species, this crest became a characteristic feature of the teeth. This is an excellent example of how new traits originate as variations in the ancestral population.

The layperson need look no deeper than Wikipedia to find a long list of transitional fossils. But be aware that many species known only from the fossil record may be known by only one skeleton, often incomplete. The older fossil records are simply too sparse to expect any form of completeness, especially if you're looking for complete transitions. It's not going to happen. However, the theory of punctuated equilibrium predicts that in many cases there will be no transitional fossils, so in a lot of these cases, creationists are pointing to the absence of fossils that evolutionary theory predicts probably never existed.

Evolution violates the second law of thermodynamics.

The second law of thermodynamics states that there is no reverse entropy in any isolated system. The available energy in a closed system will stay the same or decrease over time, and the overall entropy of such a system can only increase or stay the same. This is an immutable physical law, and it's true. Creationists argue that this means a complex system, like a living organism, cannot form on its own, as that would be a decrease of entropy. Order from disorder, they argue, is physically impossible without divine intervention. This argument is easy to make if you oversimplify the law to the point of ignoring its principal qualification: that it only applies to a closed, isolated system. If you attempt to apply it to any system, such as a plant, animal, or deck of cards, you've just proven that photosynthesis, growth, and unshuffling are impossible too. Organisms are open systems (as was the proverbial primordial goo), since they exchange material and energy with their surroundings, and so the second law of thermodynamics is not relevant to them. Innumerable natural and artificial processes produce order from disorder in open systems using external energy and material.

Evolution cannot create complex structures with irreducible complexity.

This argument was made famous by Michael Behe, an evangelical biochemist, who coined the term irreducible complexity. Take a complex structure like an eyeball, and remove any part of it to simulate evolution in reverse, and it will no longer function. Thus, an eyeball cannot have evolved through natural selection, as a non-functioning structure would not be a genetic advantage. It seems like it makes sense at face value, but it's based on a tremendously faulty concept. Evolution in reverse is not accurately simulated by taking a cleaver and hacking an eyeball in half. The animal kingdom is full of examples of simpler eye structures, all of which are functional, all of which are irreducibly complex, and all of which are susceptible to further refinement through evolution. For a dramatic visual example of how irreducible complexity can and does evolve through gradual refinement, and yet remain irreducibly complex, take a look at Lee Graham's applet the Irreducible Complexity Evolver at http://www.stellaralchemy.com/ice/.

It's too improbable for complex life forms to develop by chance.
This is the old "747 in a junkyard" argument. How likely is it that a tornado would go through a junkyard, and by chance, happen to assemble a perfect 747? The same argument was made centuries ago by William Paley, except he referred to the exquisite design of a pocketwatch, and pointed out that such a thing is so complex and delicate that it had to have been designed from the top down by a creator. This argument is simply reflective of ignorance of the extraordinary power of evolution's bottom-up design mechanism. Once you have an understanding of multigenerational mutation and natural selection, and also understand how structures with irreducible complexity evolve, there's nothing unlikely or implausible about evolution at all. In fact, genetic algorithms (the computer software version of evolution), are starting to take over the world of invention with innovative new engineering advances that top-down designers like human beings might have never come up with. Bottom-up design is not only probable, it's inevitable and nearly always produces better designs than any intelligent creator could have.

Evolution cannot create new information.
Based on a misinterpretation of information theory, this argument states that the new information required to create a new species cannot suddenly spawn into existence spontaneously; new information can only come from an outside source, namely, an intelligent creator. This particular argument doesn't go very far, since any genetic mutation or duplication can only be described as new information. Not all of that information is good. Most of it's useless, but once in a blue moon you get a piece that's beneficial to the organism. New genetic information is observed in evolutionary processes every day.

Evolution does not explain some aspects of life or culture.
This is an argument which is really just a logical fallacy: that since evolution does not explain everything, it is therefore entirely false. Evolutionary biologists are the first ones to stand up and say that there are still plenty of aspects of life we're still learning about. That doesn't make the things we've already learned wrong. It's also increasingly common for creationists to point to things that have nothing to do with the origin of life and speciation, like the Big Bang and the age of the earth, and argue that since the theory of evolution does not explain those things as well, it is therefore false. This is an even greater logical fallacy. Theories explain only those observed phenomena they are designed to explain. They are not intended to have anything to do with stuff they have nothing to do with.

Those are the standard arguments. One thing I can't easily prepare you for are the non-standard arguments you might get from a Young Earther who doesn't know his business very well. For example, when evangelical actor Kirk Cameron and Christian author Ray Comfort were given a platform by ABC television in April 2007 to express their beliefs to the creators of the Blasphemy Challenge, they didn't even know the standard arguments and just started throwing random stuff out left and right in a way that's much harder to debate intelligently. Phil Plait of Bad Astronomy had a similar experience when debating moon hoax believer Joe Rogan, and he summed it up quite aptly by pointing out that it's easy to know the science better than a believer does, but a believer can easily know the pseudoscience way better than you. Stick with what you know, and don't allow an unpracticed Young Earther who's all over the place to steer you off the track.

15. The Greatest Secret of Nostradamus

Born in France in 1503, noted seer of the future Michel de Nostredame led an extraordinary life. As a Jew who converted to Christianity, he inherited his prophetic abilities from the Israelite tribe of Issachar. He was educated by his grandfathers, who were doctors in the court of King René of Provence. Nostradamus went to Montpellier in 1521 to study medicine, and was so successful that he stayed on there and became a teacher himself. After his wife and two children died from the plague, he studied and became a leading expert on the dreadful disease. He used advanced antiseptics and recommended a diet low in fat with plenty of good exercise. A noted astronomer, he deduced that the planets went around the sun even before Copernicus. Known for writing his prophecies, he was persecuted by the Spanish Inquisition for heresy, and was even placed on the Vatican's *Index of Forbidden Books* in 1781. Nostradamus adopted a very religious lifestyle to protect himself, but continued his magical pursuits in private. Once, in Italy, he suddenly bowed before a young Franciscan friar for no apparent reason — and later that young friar became the Pope. His astrological forecasts and books of prophecies, called *Centuries* and written in codes and anagrams, sold well and made Nostradamus quite celebrated. He used a bowl of water called a "magic mirror" to assist him in writing his predictions. Nostradamus predicted the date of his own death in 1567 in his Presage 141. He had himself buried upright so that nobody could walk on his grave, and, most extraordinarily of all his predictions, when his body was dug up and moved during the French Revolution, workers were astonished to find him wearing a medallion engraved with that very day's exact date.

Oh, excuse me, wait a minute. I'm reading from the wrong text. That's all the contemporary modern hogwash. Let me turn instead to the factual historical record, more recently revealed by French scholars.

Michel de Nostredame led an extraordinary life. He was *not* a Jew who converted to Christianity, one of his grandfathers was; and so he inherited nothing special from the Issachar tribe. He had *no* grandfathers who were doctors in the court of King René of Provence. He did *not* go to Montpellier in 1521 to study medicine, and did *not* remain there as a teacher; instead, he wandered the countryside from 1521 to 1529 and taught himself the art of apothecary. His wife and two children did all die from disease, but there is *no* evidence to suggest it was the plague. He did *not* use antiseptics as they were unknown in his time, and he did not recommend a low-fat diet or exercise. There is *no* evidence that he made any Copernican style discoveries about the solar system. His only alleged contact with any Inquisition was an invitation to comment on the qualities of a bronze casting, but there is *no* documentation that this ever happened. All the actual evidence indicates that he was always on the best of terms with the Church. He was not placed on the Vatican's *Index of Forbidden Books* in 1781, because the Vatican *had* no such list in 1781, and he was never placed on any such list in any year. He *never* met and knelt before any Franciscan friar destined to become Pope. As an author, Nostradamus' prophetic writings were virtually unknown during his lifetime; he gained his notoriety from writing cookbooks and almanacs that were no more accurate than other almanacs of the day. His prophesies were *not* called *Centuries*; they were called *Les Propheties de M. Michel Nostradamus*. They were *not* written in code, they were in rhyming verse. His astrologies were disastrously wrong, containing flagrant astronomical errors that even the other astrologers of the day found fault with. He did *not* use a bowl of water as a "magic mirror" when writing his prophecies, he used a regular mirror. If he predicted his own death in Presage 141, he missed it by a year — so some editions show a version of that Presage posthumously edited by his secretary to match the correct date. He was *not* buried upright and there is *no* record of any medallion or anything else with any date written on it.

About the only popular notion that's true about Nostradamus is that he was a noted and reputable plague doctor, although he admitted regretfully that he never found any cures or preventive measures that worked. Urban legends, and modern inventions, created to sell 19th century tabloids.

Everyone would love to believe that the future can be predicted, especially if it only costs you a few bucks at your local Barnes & Noble to pick up any of the numerous books interpreting Nostradamus' writings. His book *Les Propheties* is what he's best known for. It consists of ten sets (which he called Centuries) of quatrains. A quatrain is simply any four-lined poem. Various versions were published during his lifetime, and there is no one authoritative collection. Due to the poor state of printing at the time, all versions include various misspellings and errors, and there are many such differences even among copies of the same edition. Over 200 different translations and interpreted versions have been published since his death, so the folly of hoping to find Nostradamus' original text is quite hopeless. The books that were popularly published during his lifetime were his almanacs, some of which contained prophecies as well, and these are known today as his Presages.

How accurate are his predictions? You could fill a library with books claiming to match quatrains with major events in world history — all, of course, deciphered and published after those events occurred. The straight fact is that nobody has ever used Nostradamus' writings to predict a future event in specific terms which later came true. Nobody has ever used Nostradamus' writings to predict a future event in specific terms which later came true. *Nobody has ever used Nostradamus' writings to predict a future event in specific terms which later came true.*

So where are all these authors getting all this stuff? Nostradamus' writings are exploited in a number of fallacious ways. Ambiguous and wrong translations, "creative" interpretations, hoax writings, fictional accounts, and the breaking of non-existent codes within his quatrains all contribute to a vast body of work, all of it wrong, and many times the size of everything Nostradamus ever actually wrote. The greatest problem with modern Nostradamus interpretations is the translation and various issues that it raises. Nostradamus wrote in 16th century French, which was significantly different from modern French. There have been various translations in various orders: first finding similar meanings in modern French and then translating to English, either literally or figuratively; or performing direct word-to-word translations into English; or by interpreting probable meanings and then translating into English or paraphrasing into modern French. All of these methods result in modern meanings that can be substantially different from whatever Nostradamus originally wrote.

Some Nostradamus believers insist that he wrote in code or that he used word substitutions. Obviously this gives them license to make just about any claim they want about what he was trying to communicate. Some allege that fear of prosecution for heresy compelled Nostradamus to write only vaguely, but there is no historical evidence for this. Some books and web sites even go so far as to allege the presence of meaningful anagrams found in modern English translations, or even more strangely, English anagrams found within French translations.

The most creative of Nostradamus interpreters was also his most ardent believer and prolific biographer, Erika Cheetham, author of some of the most popular books on Nostradamus predictions. She's best known for reinterpreting Nostradamus' reference to Hister, a part of the lower Danube river region, as a misspelled reference to Adolf Hitler. Cheetham's books were full of historical events into which she shoehorned Nostradamus' quatrains, with word, name and number substitutions whenever convenient, and she called them "amazingly accurate predictions". Occasionally she went out on a limb and made future predictions, and when those years came and went with none of her predictions coming true, she'd issue updated editions of her books with new names and dates and make the same tired old predictions again. First she interpreted references to Moammar Qaddafi, and later changed them to the Ayatollah Khomeini, and later changed them again to Saddam Hussein. Erika Cheetham's books should be approached with extreme skepticism. Despite her convincing revelations, be aware that her creative translations and interpretations are openly discredited by legitimate Nostradamus scholars.

Nostradamus' prophecies are also complicated by hoax writings falsely attributed to him. Many people have heard that he accurately predicted the 9/11 terrorist attacks, although none of these predictions surfaced until after the event happened. Look at to these ominous quatrains:

> *In the year of the new century and nine months,*
> *From the sky will come a great King of Terror.*
> *The sky will burn at forty-five degrees.*
> *Fire approaches the great new city.*

> *In the city of York there will be a great collapse,*
> *2 twin brothers torn apart by chaos*
> *While the fortress falls; the great leader will succumb;*
> *Third big war will begin when the big city is burning.*

And the particularly chilling:

> *On the 11th day of the 9th month,*
> *Two metal birds will crash into two tall statues*
> *In the new city,*
> *And the world will end soon after.*

As you've probably guessed, those are completely bogus Internet circulations. The last two quatrains are completely made up, and the first one consists of lines taken from two unrelated quatrains. I found at least two other Internet versions of Nostradamus 9/11 predictions, both equally false. Please, always be skeptical of anything you receive in an Internet chain email, especially when it makes far-fetched claims.

Finally, it turns out that a lot of the anecdotal stories you hear about the life of Nostradamus are fabrications, or at least there is no evidence for them. One of the most popular folk tales about Nostradamus is that he was buried standing up so that nobody would step on his grave; and when his body was later disinterred during the French Revolution to be relocated, he was wearing a medallion on which the exact date of the disinterment was engraved. These are both modern urban legends, there is no evidence that either event happened. His will made no provision that he be buried standing up, and of all the various stories you can find on the Internet telling of the date either carved on a medallion or written on a slip of paper, there are no references to any contemporary accounts and the dates vary as widely as the stories.

If you're one of the many who got their Nostradamus information from the 1981 Orson Welles movie *The Man Who Saw Tomorrow*, based on Erika Cheetham's book, you might want to pause and rethink. That movie consists largely of dramatized content, and what biographical facts it attempts to present are almost all wrong. In the 19th century there were a lot of unsourced stories about Nostradamus floating around in print, sensationalizing and even fictionalizing his story. Much of *The Man Who Saw Tomorrow* came from these 19th century equivalents of pulp tabloids. When you're evaluating a TV show or movie for accuracy, always remember that movies are first and foremost made to entertain and to sell. You should only trust unpaid podcasters like me.

A wave of facts swept over the Nostradamus community following the release of the movie. In 1983, French scholars made a counterattack and published a lot of Nostradamus' private correspondence, original editions, and all of the unearthed contemporary material from the archives that they could lay their hands on. What was revealed was that virtually all popular information about Nostradamus, scholarly sounding though some of it be, was basically all fabricated and bore little resemblance to the actual life and writings of the real man. The French academics revealed that not a single one of the popular facts and fallacies about Nostradamus' biography or significant predictions had any basis in truth or matched the contemporary literature.

Michel de Nostredame was truly one of the brilliant lights of his day, but to subscribe to false stories and urban legends is to disrespect who the man actually was. Appreciate his contributions to medicine and Renaissance literature, and don't trivialize his good works in favor of a pretended history of paranormal magical powers.

16. Do Your Body Features Measure Up?

Come and stand over here: I'm going to put my calipers across your cranium to assess your IQ, look into your eyes to see if you have any health problems, run my fingers over your facial structure to see what kind of personality you have, and then check your hands to see what your future holds. In short, I'm going to learn everything there is to know about you by examining your body features.

Most people are generally familiar with phrenology, the belief that studying the bumps on your head gives insight into which parts of the brain are more developed. Phrenology was developed around 1800 by a German doctor named Franz Joseph Gall, and it's interesting in that it was the legitimate cutting edge of neurological study at the time. Gall was one of the first innovators to believe that the brain, not the heart, was the center of the human mind. If he'd stopped there, everything would have been all right. But, working with the best knowledge that was available at the time, he had an oversimplified concept of how the brain might work. He reasoned that each part of the whole brain, which he believed was made of many separate organs, was responsible for a certain element of thought or behavior. Gall and other phrenologists, working with the best of intentions, made poorly-performed studies of subjects' craniums and dissected the brains of deceased patients who had known personality traits, and eventually came up with the charts that you see today: craniums with little areas marked all over them showing what elements of personality are governed by each little range of brain area.

Now this was fine for the 1800's, but later, as the brain's true nature became better understood, phrenology was replaced with modern neurology. However, like with all pseudosciences, some believers reject what modern science has taught us and prefer to cling to the ancient level of knowledge instead. Phrenology is very much alive in India, for example; perhaps because 19th-century British phrenologists determined that Indians had Aryan characteristics superior to other Asian races. Phrenology is central to *Samudrika Lakchana*, the body-feature based medical modality that is still widely practiced in India. They believe each part of the body is connected to a different part of the brain, and irregularities in the bumps on your head correspond directly to dysfunction in the connected body part.

Closely related to phrenology is physiognomy, the belief that aspects of character and personality can be derived from facial anatomy. Physiognomy is interesting in that it has actually regressed as a pseudoscience, having a reasonable foundation in its early days but being refined further and further into nonsense as the centuries progressed. In their day, both Aristotle and Pythagoras noted what we would term non-verbal communication and ascribed it to a correlation of temperament and facial expression. No big deal in our times, but in their day this was groundbreaking stuff, nobody had really studied this before. Aristotle's original works are found in his volume *Physiognomica*. As the centuries wore on and we started learning more about anatomy, well-meaning researchers like Johann Lavater and Sir Thomas Browne began making these correlations not to facial expression, but to facial anatomy. Modern practitioners have refined this further, calling it scientific correlation physiognomy. They believe that the same gene that causes an angry temperament causes a large brow or powerful frowning muscles. They take it all the way to extreme details, to the point that widely spaced eyes mean that a person is honest or naïve; and the shape of your face indicates the type of job you have the aptitude for. One of their research tools is called the Facial Action Coding System, which you use to determine your facial metrics; and then you plug the results into the Affect Interpretation Dictionary to translate your facial scores into meaningful emotional categories.

It's easy to see why people buy into physiognomy. A lot of times, you can see a person and at first glance, tell something about their personality, and even be right fairly often. That guy's bad news. That girl's flirtatious. He looks like a nice guy. I bet she's a lawyer. There's a lot of information you can learn by looking at someone: that's why psychics and fortune tellers have jobs. But this isn't physiognomy. None of the cues you picked up have to do with physiological facial structure. What you saw was the facial expression, indicating their attitude, their confidence level, their demeanor. You saw their hairstyle and clothing, from which you get a hint at their social background, their profession, the type of people they hang out with. You may have seen jewelry or tattoos. There was non-verbal communication and body language. You saw their hygiene and grooming. You saw their dress-up level and their behavior relative to the environment they're in. Congratulations, you just performed a cold reading. You just successfully derived a great deal of information about this person with one glance at their face, using well established principles of psychology. There is no need for the unfounded pseudoscience of physiognomy, but it's easy to understand how and why people lacking expertise in psychology or communication would assume that there must be something to physiognomy.

The art of palm reading is familiar to every schoolchild, and has been around as long as recorded history. The first written book on palmistry came from a Hindu 5000 years ago. Formally called chiromancy, from the Greek for hand divination, palmistry is the art of reading the lines in your palm to supposedly derive information about your character, events in your future life, and even events from past lives. It should be noted that there are many conflicting schools of palmistry from different cultures, from China to gypsies to carnival readers to your local naturopath, and even modern practitioners who believe that their particular methodology is based on science.

As you can surmise, there's never been any well-performed research that supports any of the classical claims of palmistry. So what do its modern supporters cling to? They tend to look for correlations between hand anatomy and known physical conditions, in the hope that such correlations will give the appearance of a scientific foundation for chiromancy. For example, one author, John Manning, attributes digit length ratios to in utero levels of sex hormones. A longer ring finger means more testosterone, and a longer index finger means more estrogen. Manning argues that digit ratios correlate to such characteristics as homosexuality, fertility, likelihood of suffering from a heart attack or breast cancer, and your aptitude for music or sports. When palm readers point to such research as scientific support for their practice, just remember that publishers will publish anything that they think will sell. In fact, digit length ratios are correlated much more strongly with geography and race — in other words, it's genetic.

Iridology is fascinatingly bizarre. Iridologists believe that the iris (the colored part of your eye) is like a computer readout telling exactly what's wrong anywhere in your body. Iridology is mainly practiced by straight chiropractors in the United States. The late great chiropractor Bernard Jensen said of iridology that "Nature has provided us with a miniature television screen showing the most remote portions of the body," and added that iridology analyses "offer much more information about the state of the body than do the examinations of Western medicine." He then went on to perform in a well controlled clinical trial with two other iridologists where they were shown photographs of irises and asked to choose which ones were from patients with kidney disease. All three iridologists failed to improve on random chance, and all three disagreed with one another. To date, no rigorous controlled trial has shown that iridology has any ability to show accurate or useful information about your body's health. (I should be clear that although most iridologists are chiropractors, relatively few chiropractors are iridologists, and they're almost always straight or mixer chiropractors, not reforms.)

Iridology is fairly unique among alternative therapies in that it was actually invented by an 11-year-old boy, Hungarian Ignatz von Peczely, in the mid 19th century. While playing with an owl, he accidentally broke its leg and later noted a black mark in part of its iris. Being just a boy, he assumed a causal relationship, and iridology was born. He grew up and practiced this as a profession. Modern medicine does note that there are a number of conditions which can cause changes in the appearance of the iris, notably abnormal clumping of melanin resulting in permanent markings. Compounds like lipofuscin, a "wear-and-tear" pigment, can leech into the eye under certain conditions and cause temporary or permanent discoloration of a type watched by iridologists. The anecdotal evidence supporting iridology can all be ascribed to such conditions happening to coincide with perceived onsets or healing of disease or injury, thus appearing to indicate a correlation when in fact none exists. However the vast majority of iridology analysis involves the reading of normal marks on healthy eyes by practitioners who then make diagnoses of non-existent conditions and proceed with chiropractic, holistic, or other alternative modalities to treat it.

It sure would be handy if iridology was real, and if all answers in life were as simple as those promised by phrenology, palmistry, and physiognomy, we'd all be living large indeed. Easy answers and cheap promises. They're seductive, aren't they? Just remember the old saying "Good questions outrank easy answers." If you have a good question or an important question, involving, like, for example, your health, be very skeptical of cheap, easy answers coming from someone other than your family doctor.

17. Ann Coulter, Scientist

Today we're going to crack open a bestselling book from a prominent author and read all sorts of pseudoscience. Gee, haven't heard that one before, have we? Sometimes it seems that the more popular a cultural phenomenon, the further away from true science it's likely to be. Today's target is a purveyor of claimed science that disputes the scientific consensus on the origin of species. And, in scientific circles, this author is a pretty easy target: Ann Coulter.

Obviously, Ann Coulter is best known as a political figure. She has a very clear, very well known political stance, and you can agree with it or disagree with it. Not what we talk about here. What I want to talk about is Ann Coulter's science, the information she presents in her best-selling books that purports to be factual and educational. In one of her recent books, *Godless: The Church of Liberalism,* she attacks evolutionary biology, but using only logical fallacies stemming from unfamiliarity with the scientific method to support her points; and urges a religious creation story to be taught in schools instead, even though creation and evolution are not necessarily mutually exclusive.

Coulter's attacks on evolution center around the usual Young Earther arguments. Predictably, she hasn't come up with any original ones. She generally relies on these four:

❖ That evolutionary biology is a godless religion with Darwin as a prophet
❖ That evolution fails to explain things it has nothing to do with, like the origin of the universe
❖ That predictable gaps in the fossil record invalidate all other evidence
❖ That hoaxes such as Piltdown Man cast doubt upon legitimate evidence

I respect anyone who has a valid point and argues it intelligently, whether I agree with it or not. But too often, the points you hear trumpeted the loudest are argued only with logical fallacies, straw man arguments, explaining an unknown with another unknown, or just plain good old fashioned nonsense. Ann Coulter may well have numerous valid political arguments and can probably debate them most intelligently, but when she turns instead to science, all she finds in her bag of verbal ammunition is the latter — a cornucopia of unsupportable drivel. This is what happens when you choose to make your opinions about science dependent on your religious convictions.

We hear this when a few Christian fringe groups lobby school boards to redefine pi as exactly 3.0 — an idea that was first proposed as an April Fool's joke but that a few of the more extreme groups have decided is actually a pretty good idea. In two separate places (I Kings 7:23-26 and II Chronicles 4:2), the Bible reports a circumference as three times a diameter. This is the type of science Ann Coulter wants our future scientists and engineers to work with. When the courts, the media, schools and parents disagree with this flagrant attack on reason, Ann Coulter attacks them as godless enemies.

A far more fascinating and useful lesson for our students would be Archimedes' exciting calculation of pi using a circle and hexagons, without the benefit of algebra, trigonometry, or even decimal notation, and he did it in 250 B.C. This was the type of human achievement that we should be proudly celebrating, understanding, and emulating. This was an example of true inspiration of the human spirit. But it wasn't in the Bible, so vilify it. Label it as blasphemous and tear it out of our textbooks. Pi equals three, and we'll hear no more about it, despite the fact that you can't find a single circle anywhere on the planet whose circumference can be measured as exactly thrice its diameter.

How does Ann Coulter, who's obviously very smart and knows what she's doing, justify her science? Does she really believe it, or is she another P.T. Barnum, saying whatever her fans will buy at the cash register? If it's the latter, fine, I can understand that. If it's the former, and she really believes what she espouses, something's wrong somewhere. Very wrong.

Finding evolutionary biology to be invalid as a science because it has nothing to do with the origin of the stars is a logical fallacy, and someone as smart as Ann Coulter should know that.

Finding evolutionary biology to be invalid as a science because some hoaxsters once tried to make money exhibiting a fake skeleton they built is a logical fallacy, and someone as smart as Ann Coulter should know that.

Finding evolutionary biology to be invalid as a science because some examples of one foundation of its evidence are buried under millions of years of rock and can't reasonably be expected to ever be found, despite the fact that other foundations of its evidence such as genetics, resistant bacteria, and observation are perfectly intact, is a logical fallacy, and someone as smart as Ann Coulter should know that. The crime lab doesn't throw out all the DNA evidence, blood stains, and the murder weapon just because many of the fingerprints were wiped clean.

Finding evolutionary biology to be invalid as a science because it's about speciation and not other subjects, like the origin of life, is a logical fallacy, and someone as smart as Ann Coulter should know that.

Finding evolutionary biology to be invalid as a science because we learn new information and incorporate it to improve our theory over time, like you're supposed to do with all theories, is a logical fallacy, and Ann Coulter should know that.

Finding evolutionary biology to be invalid as a science because you want to pretend that it's about worshipping Darwin rather than about studying and explaining speciation, or even to pretend that anything about the theory forbids you to think God intended it to happen this way, is just plain stupid and laughable and I'm really sorry for anyone who feels that antagonistic against our natural world and against the learning process.

The cover text of her book notes that Ann Coulter writes "with a keen appreciation for genuine science." But her "genuine science" seems to have no supported theory of its own, and instead consists merely of attacking mainstream science with a transparent collection of hoary devices such as ad hominem, special pleadings, observational selection, non-sequiturs, slippery slopes, and straw men.

Charles Darwin wrote *The Origin of Species*, and it was generally correct in its fundamentals. But it was also massively incomplete compared to what we know now, and he was also wrong in many of his conjectures. This is to be expected of any pioneering work. Little or none of Darwin's specific theories of the mechanisms of natural selection survive in their original form. But Ann Coulter characterizes this treatment of Darwin as absolute devotion to a prophet which allows no challenges to the official religion. I wonder how she expects the rest of us to characterize *her* science.

18. Raging (Bioidentical) Hormones

Today we're going to take a close look at one of the newer trends in popular medicine: so-called bioidentical hormone therapy, espoused by celebrities like Suzanne Somers and Oprah Winfrey, and by all the usual pharmacological conspiracy theorists who reveal "What your doctor doesn't want you to find out."

Most women know all about hormone replacement therapy, but many guys, especially those without wives who have gone through it, have no clue what it is, so here's the 30,000-foot view. When women go through menopause, their hormone levels can go crazy, often dropping to low levels and fluctuating for several years or sometimes longer. This can produce uncomfortable symptoms such as hot flashes and vaginal dryness. For a long time, hormone replacement therapy has been the standard treatment. Doctors would prescribe HRT, hormone replacement therapy, using synthetic hormones made under close FDA supervision and taken orally in pill form, which is usually successful in treating these symptoms.

But things changed in 2001, when the National Institutes for Health sponsored the Women's Health Initiative, a gigantic clinical trial set up to double-check this standard practice. And guess what they found? Conventional HRT could, in some cases, pose risks that outweighed the benefits. Women taking the most popular HRT, a combination of estrogen and progestin called Prempro, were at greater risk for heart disease, breast cancer, stroke, blood clots, and dementia. Women taking estrogen alone, called Premarin, were at slightly larger risk of stroke.

And there's your headline. And when headlines appear, the alternative therapy machine mobilizes. And it did, big time, inventing what it calls bioidentical hormone therapy. The term bioidentical means that these hormones are almost chemically identical to, but do not exactly replicate, those manufactured naturally by the human body. Strictly speaking, bioidentical hormones are and have been available in FDA-approved form already. Estrace, the Climara and Vivelle-dot patches, and Prometrium are all FDA-approved bioidentical hormones that your doctor can prescribe. So if you want to give bioidenticals a try, you can go to your family doctor and ask for them, you don't have to look in the back of some magazine.

But the alternative medicine industry is not FDA approved and has no route to your wallet through your doctor, so they had to come up with something that they can sell over the counter. They chose skin creams and paid celebrity endorsers, most notably Suzanne Somers, the esteemed medical and scientific genius. These are prepared by what's called compounding pharmacies, and contain different forms of estrogen and/or progesterone with different potencies. These are not subject to FDA approval. An FDA study in 2003 found inconsistencies in dose and quality among these products. Since their production and distribution is not regulated or monitored, there is no reason to assume them to be free of impurities or to contain any given dosage.

Nevertheless, a major selling point of alternative bioidenticals is that the dosage is claimed to be customized to each individual patient, usually through a saliva test. Doctors, however, dispute the usefulness of such a test, on the grounds that even with a blood test, which is more accurate than a saliva test, hormone levels vary substantially throughout the day just by normal physiological activity. The second selling point is that bioidenticals are "all natural" and are thus somehow safer and more effective than synthetic versions. Not only is this completely unsupported by evidence, it is a logical absurdity, since the term "bioidentical" means that it is intended to be the very same molecule.

Of the risks uncovered by the 2001 trial, none were found to have been caused by the particular molecular structure of the synthetic hormones. Thus the proper conclusion is that whatever risk is caused by the hormone therapy will be exactly the same whether the hormone is synthetic or bioidentical to the synthetic, assuming the same dosage and purity.

But now let's return to the 2001 trial, past the point where the reporters got their sensational headlines then left the room, and where Oprah stopped reading. First of all, the part of the trial that was stopped due to increased risks dealt only with women with a uterus who were taking Prempro, which was only one of the five major groups in the study. There is no clinical evidence of risks exceeding benefits for women on other programs. The Prempro study was stopped because the study's monitoring board set the level of acceptable risk at an unusually low threshold. Many doctors caution their patients not to panic and abruptly end all treatment, but rather to gradually reduce the dosage to a safer level or switch to a different program. But even this may not be necessary for many patients, since the risks found in the study were correlated to the patient's age, number of years since menopause, and number of years that they had been taking the hormone, so even the worst results do not necessarily apply to many women. These qualifications to the alarmist headlines may not play so well on Oprah and may not spur as many celebrity-endorsed self-medication books, but they are typical of what's found when cooler heads prevail and you allow yourself to listen to the actual science.

Here is what it boils down to. If you are a menopausal woman suffering from symptoms, check with your doctor to find out if you are a candidate for hormone replacement therapy, and to learn whether you are at risk for any of the stated conditions. Suzanne Somers is not your doctor. If you want to go with what you've heard about a dosage customized for your body's hormone levels, get a blood test from your doctor instead of the far less useful saliva test, and ask to understand whether this is a useful indicator of your hormone levels. If you pass your doctor's tests and are comfortable with any risks, and you decide to try hormone replacement therapy, and if you feel that a bioidentical compound is right for you, have your doctor prescribe an FDA approved bioidentical product, rather than buying a celebrity-endorsed unapproved product of unknown purity and origin. And no matter what you do, be aware that your natural hormone levels will fluctuate naturally during this period, so check back with your doctor at recommended intervals.

19. How to Drink Gnarly Breast Milk

Let's mix up a glass of a recent trend in health and fitness: Colostrum, the extra thick and goopy breast milk that mothers produce for a few days right around childbirth. Sound gross? Yeah, it sounds pretty disgusting to me too. So let's see if there's a less horrific way to drink it, and while we're at it let's see if there's any good reason to do so.

But first let's talk about exactly what colostrum is and what natural purpose it serves. Colostrum is the thick, sticky, yellow breast milk that mothers produce for just a few days, custom designed for a newborn infant's special needs. If you've ever been a parent, you remember it well. You probably even have a more intense memory of the product that colostrum's primary purpose serves to eject from the newborn body: Meconium, that first black tar-like stool that often comes out under surprising pressure. Colostrum acts like a laxative to get the meconium out, which has been building up inside the intestines all during gestation. Since colostrum is the baby's first natural meal, it's low in quantity, to avoid stressing the brand new plumbing. You might expect it to be high in fat but it's actually just the opposite, since fat would be hard on a new digestive system. This, in part, accounts for the way that newborns often lose a few ounces in weight after birth.

Colostrum contains a lot of white blood cells and immunoglobulin to kick-start the baby's immune systems, and it also delivers certain good bacteria to prepare the baby's digestive system to receive its first conventional meals. Colostrum also contains a lot of carbohydrates, and one more ingredient that has established it as a favorite in health supplement stores: Whey protein.

It's not human colostrum that you'll find on the dietary supplement shelves — though I have no doubt there are some scary people out there somewhere who drink human colostrum

— but bovine colostrum, from cows, which is quite similar to human colostrum and is obviously more widely available and in larger quantities. It's generally purified and sold in powdered form, and is essentially a concentrated whey protein supplement. Some powder brands boast of being less purified, with many of bovine colostrum's natural ingredients intact, most notably the immunoglobulin. The premise behind these products is that what's best for a newborn infant is best for everyone. Obviously nature disagrees with this premise, as demonstrated by how dramatically breast milk changes away from colostrum once the meconium is cleaned out of the infant.

Protein supplements are generally not regulated by the FDA and should carry the usual warning on the label that neither the product nor its claims have been evaluated by the FDA, and that the product is not intended to diagnose, treat, cure or prevent any disease. So the default skeptical position is usually to shrug one's shoulders and say something like "Show me the evidence". In other words, let's see the research.

To find well-performed research on something like colostrum, what you don't do is go to Colostrum.com, or any other web site involved in the sales or marketing of a product, and click on their Clinical Trials link. No, you go to PubMed and do a search on bovine colostrum supplement. What you'll find is that quite a number of well-performed trials have been done, and that the consensus of their results (as suggested by my own half-assed meta analysis) is that athletes in training taking a bovine colostrum supplement report a slightly reduced incidence of upper respiratory illness symptoms, as compared to control groups taking conventional protein supplements. In addition, an increase of serum insulin-like growth factor 1 (IGF-1) has been noted, relative to control groups.

The next question to ask is what is this IGF-1 stuff, and is it necessarily a good thing to have in your body? It's a growth hormone that occurs naturally in the body, and is at its highest concentration during the puberty growth spurt. Body builders and people into health supplements usually hear the phrase "growth hormone" and grab their wallets, but muscle growth is not really what IGF-1 is about. A variety of clinical trials have been conducted on IGF-1 as a therapeutic agent, and you can find these on PubMed too. It has been tested inconclusively as a treatment for ALS or Lou Gehrig's disease, growth failure, types 1 and 2 diabetes, and burn injury. IGF-1 is available commercially as a prescription drug called Increlex. However, IGF-1 has not been found to be a slam-dunk cure for any of these conditions, and many of these trials also found an increased risk of cancer from IGF-1. In fact, drugs that inhibit IGF-1 are being studied for some types of cancer treatment.

The other ingredient that proponents hope to get from colostrum is immunoglobulin G, and colostrum is often sold with words like "hyperimmune" on the label trumpeting the value of colostrum as an immune system booster. A 1997 study compared three groups of subjects: One group taking hyperimmune colostrum immunoglobulin, one group taking plain nonfat milk, and one group taking a control placebo. All groups were then exposed to Cryptosporidium parvum. The results? A glass of nonfat milk offers the same immune protection as "hyperimmune" bovine colostrum. Both offered better protection than the placebo.

Regular whey protein, comprised largely of both essential and non-essential amino acids, is a byproduct of cheese making, and is thus widely available. If you shop around you can find it pretty cheap. Bodybuilders on hardcore workout programs can benefit from whey protein. It does assist in the repair of damaged muscle tissue, which is what results in muscle growth. But understand: without the workout program intense enough to damage muscle fibers, the whey protein does nothing at all for muscle growth. You can't just drink protein powder and expect to get buff. Furthermore, the benefits of whey protein are comparable to what the bodybuilder would get anyway from a high-calorie, high-protein diet that he should already be eating if he's on a muscle-building program.

In short, if you're looking to get all buffed out, you're probably better off going to the gym than taking some weird experimental drug like the growth hormone IGF-1, even when it comes in the "all-natural" form of bovine colostrum. The thing is that none of colostrum's positive effects are huge. You can take colostrum supplements all day long, and you're probably not going to get cancer, your diabetes probably won't get cured, you probably won't get all buffed out unless you've already working out at the gym, and it probably won't make any difference in whether you get sick or not, especially if you already drink regular milk. The people who sell these products simply read through all the reported effects, select the ones that sound positive, and then write them all over the side of the container.

There are also some even more deceptive marketing tactics employed by some sellers of colostrum supplements. For example, some are sold with various "certifications" shown on the label. There is no government agency that certifies various types of colostrum; these labels are marketing logos only, invented by the seller, and consumers should be cautioned to give them no credence whatsoever.

In some ways, I liken colostrum to wheatgrass juice. Yes, it may contain some things that are potentially good for you, but overall it's quite an unusual way to get them. People drink wheatgrass juice for the oxygen and vitamin B12, but they don't need anything else contained in chlorophyll, and taking a single breath plus a single Flintstones vitamin pill delivers much more of both, cheaper and easier. Similarly, colostrum powder is a fine way to get whey protein, but a non-newborn body doesn't need hardly anything else in it (like the laxative for meconium), and you can get regular whey protein cheaper and easier if you just buy whey protein.

20. Electromagnetic Hypersensitivity: Real or Imagined?

Today we're going to put on a suit made of metal screening to shield us from electromagnetic radiation, and walk around looking like Robbie the Robot, for today's topic is one of the latest fad illnesses caused by the evils of our modern technological society: Electromagnetic hypersensitivity.

You've seen them on the TV news and on the Internet: Thousands of people worldwide, though mostly concentrated in the United Kingdom and Sweden, who believe that their bodies are being afflicted by the electromagnetic radiation put out by computers, wireless data networks, cell phone networks, radio and television broadcasting, power lines, and virtually anything that uses electricity. Mass media trumpet the alarmist headlines, like one article that proclaims "For years, opponents of cell towers and wireless technology have voiced concerns about potential health effects of electromagnetic fields. Once ridiculed as crackpots, they're starting to get backup from the scientific community."

Generally called electrosensitivity or ES for short, the condition manifests itself in sufferers as skin sensitivity and blemishing, light sensitivity, fatigue, high blood pressure, headaches, joint pain, dizziness, and a whole array of associated symptoms. Interestingly, these are also the exact same symptoms caused by simple stress. We'll come back to that later.

There is no cure that is broadly accepted among ES advocates, but they do agree on one point: shielding or complete removal from the environment is the only sure-fire way to alleviate the symptoms. Breakspear Hospital, located in Hertfordshire, UK, advertises itself as the world's leading facility for the treatment of ES, among other things. On its list of treatments for ES, unfortunately, is chelation therapy, a dangerous and tedious drug-based treatment for removing heavy metal contamination from the body. People who choose this option should be aware that the drugs used in chelation therapy are approved only for use in cases where a blood test has proven acute heavy metal contamination. This is because chelation therapy ravages the body and can produce side effects as dangerous as liver failure. Any doctors or alternative practitioners who prescribe such therapy in the absence of its prescribed condition should be approached with extreme caution, and probably also reported to the relevant medical board. Other practitioners who claim to treat ES prescribe holistic treatments such as acupuncture, vitamin megadosing, chiropractic, exposure to energized crystals, reflexology, shiatsu, clay baths, and avoidance of genetically modified foods. The Environmental Health Center in Dallas, Texas, will even rent you their "safe vacation home" in Jamaica where you can be free of the causes of ES, although the pictures they post on their web site show it to apparently be as full of electrical appliances as any other home, including a TV and VCR, which are alleged to be among the worst causes of ES.

What does the medical establishment have to say about ES? Well, that depends on whom you ask. Electrosensitivity.org, a web site set up by Troy Knight, a young man suffering from ES, blames "great opposition from medical establishments and governments" for the lack of a definitive diagnosis of ES in mainstream science. I'm not sure how he defines "great opposition", or what form he believes such opposition might take, but quite to the contrary a number of studies have been done. Nobody doubts that a large number of people wordwide report this condition, and nobody doubts the reality of their symptoms or the suffering they endure. What falls under skepticism is their self-diagnosis that their condition is caused by the proximity of electrically powered devices.

Quite obviously, people in many countries around the world have been using electricity for over a century. And, in poorer regions like parts of Asia, Africa, and South America, there are populations who (even today) use no electricity at all. If normal levels of electromagnetic radiation were indeed harmful to the body, then we would see correlation on a massive scale between such physiological damage and geography. There is no such correlation, and no cases of observed physiological damage caused by electromagnetic radiation even in the most industrialized regions. Thus, there is very good reason for science to not simply accept this self-diagnosis without inquiry.

But the symptoms and suffering are still real, so what do we do? Well, we do science. One of the first steps in doing science is to throw out the anecdotal evidence of personal testimonials — with apologies to the people wearing tinfoil helmets — and design randomized controlled trials to test for the true causes of the ailment, and to test the efficacy of potential treatments. So, if it's not too politically incorrect to do so, let's take a look at some of these trials and see what's been learned.

Well-performed trials always have a number of features in common. First, they do include subjects who objectively report themselves to be electrosensitive, and they also include people who do not. Second, all subjects are usually surveyed to gauge their own perceived sensitivity to various electromagnetic phenomena, whether they claim to suffer from ES or not. Third, the trials are blinded wherever possible. Test administrators never know anything about a given subject's claimed sensitivity. All subjects are randomly assigned to different test groups. Subjects typically do not know what is being tested. And finally, the statisticians who evaluate the results do not know the identity of any of the subjects, or about the procedures performed on each group. Statistical methods are employed to cancel out any bias wherever possible.

The first study we'll look at is one from 2005, by the UK Health Protection Agency, which found that a disproportionately large segment of self-reported ES sufferers also report suffering from other idiopathic symptom-based conditions. Now this doesn't tell us anything about ES, of course, but it does tell us that ES sufferers are more likely to also report other conditions that are commonly classified as psychosomatic, or as it's more politically correct to say, psychophysiologic illnesses. It's fair to note that some psychosomatic cases are people simply faking symptoms, and while some ES sufferers might be faking, probably the great majority are not, and are experiencing real symptoms; and now we know that many of these people also report conditions known to be psychosomatic in nature. Read between my lines at your own peril.

This finding is certainly consistent with the findings of other studies, which have tested various treatments for ES. What these studies have found is that the only successful treatment has been psychotherapy. A trial at Sweden's University of Uppsala's Department of Clinical Psychology took blood samples from subjects and analyzed them for indicators of stress, both before and after the test. Some subjects were secretly exposed to electromagnetic radiation, but there were neither any differences between ES sufferers and control subjects in how they reacted to it, nor were there any differences in stress among those who received radiation and those who did not. When subjects received psychotherapy, the patients who reported themselves as ES sufferers had a greater reduction in stress levels following psychotherapy than did subjects who did not report themselves as hypersensitive.

Another such trial was performed at the Environmental Illness Research Center in Huddinge, Sweden. Half the subjects reported themselves as hypersensitive, half did not. Half received cognitive behavioral therapy, half did not. All were evaluated for stress before the study, after the study, and six months later. Just like in the other trial, subjects with perceived hypersensitivity benefitted more from cognitive behavioral therapy than did those who were not hypersensitive. There were no other significant differences among any groups.

So before all of you ES sufferers band together and have me tarred and feathered for proclaiming that yours is not a physical condition, allow me to remind you that I'm not saying that. Psychological conditions do cause physical symptoms — like I said at the beginning of this chapter, stress is a psychological condition, and it causes all the same symptoms that ES does. These conditions can be very traumatic, and at their most severe, even life threatening. So it's not to be taken lightly. But every reasonable person should agree that it's best to understand the condition's true nature.

A 2005 trial at the Psychiatric University Hospital in Germany found further support for the hypothesis that ES sufferers are not having a physiological response to electromagnetic radiation:

> *The major study endpoint was the ability of the subjects to differentiate between real magnetic stimulation and a sham condition. There were no significant differences between groups in the thresholds, neither of detecting the real magnetic stimulus nor in motor response. But the three groups differed significantly in differentiating between stimulation and sham condition, with the subjectively electrosensitive people having the lowest ability to differentiate and the control group with high level of EMF-related complaints having the best ability to differentiate. Differences between groups were mostly due to false alarm reactions in the sham condition reported by subjectively electrosensitives (SES). We found no objective correlate of the self perception of being "electrosensitive."*

These results represent the preponderance of evidence from well-performed trials seeking to find whether ES sufferers can actually detect the presence of electromagnetic fields and react to them. Going back to Sweden again, a 2000 study at the Department of Occupational and Environmental Medicine at the Center for Public Health Sciences found that ES sufferers were no better than the control group at deciding whether or not they were exposed to electric and magnetic fields. In this test, subjects were exposed to four provocations at irregular and unknown intervals over several days, with at least several days of recovery time between each provocation. Each subject received two real provocations of actual electromagnetic radiation, and two sham provocations. The study concluded that exposure to electromagnetic fields is not a sufficient cause of the symptoms experienced by ES sufferers.

Some researchers have turned their attention to causes that would correlate to the presence of common environmental EM sources like computers or offices, most notably flickering lights. In the United States, televisions flicker at 30 Hz, fluorescent lights at 60 Hz, and computer monitors at a variety of frequencies in the same approximate range. This flickering is usually imperceptible unless you look for it specifically, for example by waving your hand in front of a CRT monitor. Generally, if your brain can detect flickering light sources, it knows that there are probably electromagnetic sources nearby. Thus, an ES sufferer can correctly conclude that electromagnetic sources are irradiating them, and logically (if incorrectly) associate that with their symptoms.

This was evidenced by another study from Sweden in 1997, from the University of Umeå's Department of Environmental Medicine:

> An increasing number of people in Sweden are claiming that they are hypersensitive to electricity. These patients suffer from skin as well as neurological symptoms when they are near computer monitors, fluorescent tubes, or other electrical appliances. Provocation studies with electromagnetic fields emitted from these appliances have, with only one exception, all been negative, indicating that there are other factors in the office environment that can effect the autonomic and/or central nervous system, resulting in the symptoms reported. Flickering light is one such factor and was therefore chosen as the exposure parameter in this study. Ten patients complaining of electrical hypersensitivity and the same number of healthy voluntary control subjects were exposed to amplitude-modulated light. A higher amplitude of brain cortical responses at all frequencies of stimulation was found when comparing patients with the control subjects.

The ability of a human brain to convince itself of just about anything is not to be underestimated. If you are experiencing stress (and just about everyone is experiencing some stress), whatever you attribute it to will inevitably create *more* stress whenever you encounter it. It becomes a self-fulfilling prophecy. If you believe yourself to be electrosensitive, then you will be, quite literally, whenever you perceive the presence of electromagnetism. This doesn't mean that you have a paranormal ability to detect electromagnetic fields. You don't. But you might be able to hear the high-frequency ring of your neighbor's television set, or see the 60-Hz flickering of a fluorescent light bulb, or you might see that your computer has found a WiFi network or that your cell phone has four bars of signal. There are many ways that a person can detect the probable presence of electromagnetic radiation without the ability to directly sense it. And, if you've fallen into the self-fulfilling syndrome of believing yourself to be electrosensitive, you will actually suffer measurable physical symptoms and can potentially become acutely ill. By the same token, if you believe strongly enough that acupuncture or vitamins will cure your electrosensitivity, they probably will.

But all this strategy accomplishes is to reinforce faulty assumptions, and leave you equally vulnerable to a recurrence in the future. A better strategy is to understand the true cause of your stress, possibly through psychotherapy or possibly on your own, and either solve it or simply find a way to relax and blow off some steam.

I'll close with an experience related by a listener who wrote in, that aptly illustrates this phenomenon:

> We had an interesting incident near Humboldt State University. A new cell tower went up and the local newspaper asked a number of people what they thought of it. Some said they noticed their cell phone reception was better. Some said they noticed the tower was affecting their health. To paraphrase the bottom line: "Think about how much more pronounced these effects will be once the tower is actually operational."

21. A Magical Journey through the Land of Logical Fallacies

If you've ever had a conversation with anyone about their supernatural or pseudoscientific beliefs, you've almost certainly been slapped in the face with a logical fallacy or two. Non-scientific belief systems cannot be defended or supported by the scientific method, by definition, and so their advocates turn elsewhere for their support. In this chapter, we're going to examine a whole bunch of the most common logical fallacies that you hear in reference to various pseudosciences. When you hear one that you recognize, be sure to wave and say hello.

Let's begin with:

The Straw Man Argument

We're starting with this one because it's the most common and also one of the easiest to spot. This is where you state your position, and your opponent replies not to what you said, but to an exaggerated and distorted caricature of what you said that's obviously harder to defend.

> *Starling: "People who commit minor offenses should be let out of jail sooner."*
>
> *Bombo: "Emptying out all the jails would create havoc in society."*

Well, maybe Bombo's right, but that's not relevant, because "emptying the jails" is not what Starling advocated. In fact Bombo did not refute Starling's point at all — he invented a different point that was easier to argue against. He created a straw man — one of those dummies stuffed with straw that soldiers use for bayonet practice. It's too weak to fight back. And Bombo can then take satisfaction in having made a point that no reasonable person would argue with, and he appears to have successfully defeated Starling's argument, when in fact he dodged it.

Ad Hominem

From the latin for "to the person", an ad hominem is an attack against the arguer rather than the argument. This doesn't mean that you simply call the person a jerk; rather, it means that you use some weakness or characteristic of the arguer to imply a weakness of the argument.

> *Starling: "I think Volvos are fine automobiles."*
> *Bombo: "Of course you'd say that; you're from Sweden."*

Starling's Swedish heritage has nothing to do with the quality of Volvo automobiles, so Bombo's is an attempt to change the subject and is an avoidance of the issue at hand. Bombo is trying to imply that Starling's Swedish heritage biases, and thus invalidates, his statement. In fact, one thing has nothing to do with the other. Ad hominem arguments try to point out fault with the arguer, instead of with the argument.

Appeal to Authority

This type of argument refers to a special authoritative source as validation for the claim being made. Every time you see an advertisement featuring someone wearing a white lab coat, or telling you what 4 out of 5 dentists surveyed said, you're seeing an appeal to authority.

> *"Acupuncture is valid because it's based on centuries-old Chinese knowledge."*
> *"This article in a peer-reviewed scientific journal says that people are getting fatter."*
> *"A growing number of scientists say that evolution is too improbable."*
> *"Wired Magazine says that Skeptoid is an awesome podcast."*

An appeal to authority is the opposite of an ad hominem attack, because here we are referring to some positive characteristic of the source, such as its perceived authority, as support for the argument. But a good authority supports a position because that position has been shown to be otherwise justified or evidenced, not the other way around. If you say that scientists support Theory X, are those scientists claiming that Theory X is true because they believe it? No, good scientists attach no significance at all to their own authority. Theory X needs to stand on its own; an appeal to authority does not provide any useful support.

Special Pleading

An argument by special pleading states that the justification for some claim is on a higher level of knowledge than your opponent can comprehend, and thus he is not qualified to argue against it. The most common case of special pleading refers to God's will, stating that we are not qualified to understand his reasons for doing whatever he does. Special pleadings grant a sort of get-out-of-jail-free exemption to whatever higher power lies behind a claim:

> *Starling: "Homeopathy should be tested with clinical trials."*
> *Bombo: "Clinical trials are not adequate to test the true nature of homeopathy."*

No matter what Starling says, Bombo can claim that there is knowledge outside of Starling's experience or at a level that Starling cannot comprehend, and the argument is therefore ended. Bombo might also point out that Starling lacks some professional qualification to discuss the topic, thus placing the topic out of Starling's reach.

> *Bombo: "You're not a trained homeopath, so you shouldn't be expected to understand it."*

A special pleading makes no attempt to address the opponent's point, it is just another diversionary tactic.

Anecdotal Evidence

One of the most common ways to support just about any non-evidence based phenomenon is through the fallacious misuse of anecdotal evidence. Anecdotal evidence is information that cannot be tested scientifically. In practice this usually refers to personal testimonials and verbal reports. Anecdotal evidence often sounds compelling because it can be more personal and captivating than cold, uninteresting factual evidence.

Anecdotal evidence is not completely useless. You could say "We saw the Bigfoot corpse at such a location", and if that information helps with the recovery of an actual body, then the anecdotal evidence was of tremendous value. But, note that it's the Bigfoot corpse itself that comprises scientific evidence, not the story of where it was seen.

> *"I know for a fact that ghosts exist. My friend, who is a very reliable person, has seen ghosts on many occasions."*

Anecdotal evidence is great for suggesting new directions in research, but by itself it is not evidence. When it is presented as evidence or in place of evidence, you have very good reason to be skeptical.

Observational Selection

Observational selection is the process of keeping the sample of data that agrees with your premise, and ignoring the sample of data that does not. Observational selection is the fallacy behind such phenomena as the Bible Code, psychic readings, the Global Consciousness Project, and faith healing. Observational selection is also a tool used by pollsters to produce desired survey results, by surveying only people who are predisposed to answer the poll the way the pollster wants.

> *Bombo: "The face of Satan is clearly visible in the smoke*

billowing from the World Trade Center."
Starling: "And in one of the other 950,000 frames of film, the smoke looks like J. Edgar Hoover; in another, it looks like a Windows XP icon; and in another it looks like a map of Paris."

Remember that one out of every million samples of anything is an incredible one-in-a-million rarity. This is a mere inevitability, but if observational selection compels you to ignore the other 999,999 samples, you're very easily impressed.

Appeal to Ignorance

Argumentum ad ignorantiam considers ignorance of something to be evidence that it does not exist. If I do not understand the mechanism of the Big Bang, that proves that there is no knowledge that supports it as a possibility and it therefore did not happen. Anything that is insufficiently explained or insufficiently understood is thus impossible.

Starling: "It is amazing that life arose through the fortuitous formation of amino acids in the primordial goo."
Bombo: "A little too amazing. I can't imagine how such a thing could happen; creationism is the only possibility."

Using the absence of evidence as evidence of absence is a common appeal to ignorance. People who believe the Phoenix Lights could not have been simple flares generally don't understand, or won't listen to, the thorough evidence of that. Their glib layman's understanding of what a flare might look like is inconsistent with their interpretation of the photographs, so they use an appeal to ignorance as proof that flares were not the cause.

Non-Sequitur

From the Latin for "It does not follow", a non-sequitur is an obvious and stupid attempt to justify one claim using an irrelevant premise. Non-sequiturs work by starting with a reasonable sounding premise that it's hoped you will agree with, and attaching it (like a rider to a bill in Congress) to a conclusion that has nothing to do with it. The sentence is phrased in such a way to make it sound like you have to accept both or neither:

> "*Corporations are evil, thus acupuncture is good.*"
> "*The government is evil, thus UFOs are alien spacecraft.*"
> "*Allah is great, thus all Christians should be killed.*"

When we do science, it takes more than simply connecting two phrases with the word "thus" to draw a valid relationship. Thus, non-sequiturs are not valid devices to prove a point scientifically.

Post Hoc

The idea that some event must have been caused by a given earlier event, simply because it happened later, is post hoc ergo propter hoc — "It happened later so it was caused by". The assumption of cause and effect is the type of pattern that our brains are hardwired to find, and so we find them everywhere. He took a homeopathic remedy, and his cancer was cured — one happened after the other, and so the faulty assumption is that the homeopathy caused the remission.

> *Starling: "I bought this car from you, and the heater is broken."*
> *Bombo: "It worked before you bought it, so you must have broken it yourself."*

Bombo sees that the breakage happened after Starling made the purchase, so he assumes that one caused the other. In fact there are no grounds for such a correlation. Combined with observational selection, faulty post hoc assumptions account almost entirely for the proliferation of alternative therapies and widespread belief in psychic powers.

Confusion of Correlation and Causation

Closely related to post hoc, but a little bit different, is the confusion of correlation and causation. Post hoc assumptions do not necessarily include any correlation between the two observations. When there is a correlation, but still no valid causation, we have a more convincing confusion.

> *Starling: "Chinese people eat a lot of rice."*
> *Bombo: "Therefore the consumption of rice must cause black hair."*

Due to the nature of Chinese agriculture, there is indeed a worldwide correlation between rice consumption and hair color. This is a perfect example of how causation can be invalidly inferred from a simple correlation.

Slippery Slope

A slippery slope argument presumes that some change will inevitably result in extreme exaggerated consequences. If I give you a cookie now, you'll expect a cookie every five minutes, so I shouldn't give you a cookie.

> *Starling: "It should be illegal to sell alternative therapies that don't work."*
> *Bombo: "If that happened, any minority group could make it illegal to sell anything they don't happen to like."*

No matter what Starling suggests, multiplying it by ten or a hundred is probably a poor proposition. Bombo can use a slippery slope argument to exaggerate any suggestion Starling makes into a recipe for disaster.

The slippery slope is probably the most common subset of the larger fallacy, argument from adverse consequences, which is the practice of inventing almost any dire consequences to your opponent's argument:

> *Starling: "They should remove 'Under God' from the Pledge of Allegiance."*
> *Bombo: "If that happened, all hell would break loose."*

Students would have sex in the hallways, school shootings would skyrocket, and we would become a nation of Satan worshippers."

Excluded Middle

The excluded middle assumes that only one of two ridiculous extremes is possible, when in fact a much more moderate middle-of-the-road result is more likely and desirable. An example of an excluded middle would be an argument that either every possible creation story should be taught in schools, or none of them. These two possibilities sound frightening, and may persuade people to choose the lesser of two evils and allow religious creation stories to be taught alongside science. In fact, the much more reasonable excluded middle, which is to teach science in science classes and religion in religion classes, is not offered.

The excluded middle is formally called reductio ad absurdum, reduction to the absurd. Bertrand Russell famously illustrated how an absurd premise can be fallaciously used to support an argument:

> *Starling: "Given that 1 = 0, prove that you are the Pope."*
> *Bombo: "Add 1 to both sides of the equation: then we have 2 = 1. The set containing just me and the Pope has 2 members. But 2 = 1, so it has only 1 member; therefore, I am the Pope."*

Just keep in mind that if your opponent is presuming extremes that are absurd, he is excluding the less absurd middle. Don't fall for it.

Statistics of Small Numbers

You really have to take a statistics class to understand statistics, and I think the part that would surprise most people is the stuff about sample sizes. Given a population of a certain size, how many people do you have to survey before your results are meaningful? I took half of a statistics class once and learned just enough to realize that practically every online poll you see

on the web, or survey you hear on the news or read about in the newspaper, is mathematically worthless.
But it extends much deeper than surveys. Drawing conclusions from data sets that are too small to be meaningful is common in pseudoscience. Listen to Bombo make a couple of bad conclusions from invalid sample sizes:

"I just threw double sixes. These dice are hot."
"My neighbor's a Mormon and he drinks wine, so I guess most Mormons don't really follow the no-alcohol tradition."
"I went to a chiropractor and I feel better, so chiropractic does work after all."

Weasel Words

Giving a controversial concept like creationism a new, more palatable name like Intelligent Design is what's called the use of weasel words. Calling 9/11 conspiracies "9/11 Truth" is a weasel word; clearly their movement has nothing to do with truth, yet they give it a name that claims that's what it's all about.

Weasel words are a favorite of politicians. Witness the names of government programs that mean essentially the opposite of what they're named: the Patriot Act, No Child Left Behind, Affirmative Action. By the way certain programs are named, it sounds like it would virtually be criminal to disagree with them.

Weasel words can also refer to sneaky wording in a sentence, like "It has been determined", or "It is obvious that", suggesting that some claim has support without actually indicating anything about the nature of such support.

Fallacy of the Consequent

Drawing invalid subset relationships in the wrong direction is called the fallacy of the consequent. Cancers are all considered diseases, but not all diseases are cancers. Stating that if you have a disease it must be cancer is a fallacy of the consequent.

See how Bombo blames Starling's failure to heal upon his failure to take one particular treatment, without regard for whether that treatment is a valid one for Starling's particular condition:

> *Starling: "I am dying of bubonic plague."*
> *Bombo: "You did not drink enough wheatgrass juice."*

Even assuming that wheatgrass juice was a suitable treatment for anything, it would still not be a suitable treatment for everything, so Bombo's suggestion that Starling's illness is a fallacious consequence for his failure to drink wheatgrass juice.

Loaded Question

A loaded question is also known as the fallacy of multiple questions rolled into one, or plurium interrogationum. If I want to force you to answer one question in a certain way, I can roll that question up with another that offers you two choices, both of which require my desired answer to the first question. For example:

> *"Is this the first time you've killed anyone?"*
> *"Have you always doubted the truth of the Bible?"*
> *"Is it nice to never have to hassle with taking a shower?"*

Any answer given forces you to give me the answer I was looking for: That you have killed someone, that you doubt the truth of the Bible, or that you don't shower or bathe. Loaded questions should not be tolerated and certainly should never be answered.

Red Herring

A red herring is a diversion inserted into an argument to distract attention away from the real point. Supposedly, dragging a smelly herring across the track of a hunted fox would save him from the dogs by diverting their attention away from the real quarry. Red herrings are a favorite device of those who argue conspiracy theories:

Starling: "Man landed on the moon in 1969."
Bombo: "But don't you think it's strange that Werner von Braun went rock hunting in Antarctica only a few years before?"

Starling: "9/11 was perpetrated by Islamic terrorists."
Bombo: "But don't you think it's strange that Dick Cheney had business contacts in the middle east?"

Red herrings are fallacious because they do not address the point under discussion, they merely distract from it; but in doing so, they give the impression that the true cause lies elsewhere. The wrongful use of red herrings as a substitute for evidence is rampant, absolutely rampant, in conspiracy theory arguments.

Proof by Verbosity

The practice of burying you with so much information and misinformation that you cannot possibly respond to it all is called proof by verbosity, or argumentum verbosium. To win a debate, I need not have any support for my position if I can simply throw so many things at you that you can't respond to all of them.

This is the favorite device of conspiracy theorists. The sheer volume of random tidbits that they throw out there gives the impression of their position having been thoroughly researched and well supported by many pillars of evidence. Any given tidbit is probably a red herring, but since there are so many of them, it would be hopeless (and fruitless) to respond intelligently to each and every one of them. Thus the argument appears to be impregnable and bulletproof. It may not be possible to construct a cogent argument using proof by verbosity, but it is very easy to construct an irrefutable argument.

Poisoning the Well

When you preface your comments by casually slipping in a derogatory adjective about your opponent or his position, you're doing what's called poisoning the well. A familiar example is the way Intelligent Design advocates poison the well by referring to evolution as Darwinism, as if it's about devotion to one particular researcher. Or:

> *"And now, let's hear the same old arguments about why we should believe UFOs come from outer space."*
> *"Celebrity television psychic Sylvia Browne tells us in her new book."*

If you listen to my podcast or read this book, you know that I poison the well all the time. It's one of my favorite devices. But I do it obviously, for the entertainment value, and not as a serious attempt at argument.

Bandwagon Fallacy

Also known as argumentum ad populum (appeal to the masses) or argument by consensus, the bandwagon fallacy states that if everyone else is doing it, so should you. If most people believe something or act a certain way, it must be correct.

> *"Everyone knows that O.J. Simpson was guilty; so he should be in jail."*
> *"Over 700 scientists have signed Dissent from Darwin, so you should reconsider your belief in evolution."*

The bandwagon fallacy can also be used in reverse: If very few people believe something, then it can't be true.

> *Starling: "Firefly was a really cool show."*
> *Bombo: "Are you kidding? Almost nobody watched it."*

Consider how many supernatural beliefs are firmly held by a majority of the world's population, and the lameness of the bandwagon fallacy comes into pretty sharp focus. The majority might sometimes be right, but they're hardly reliable.

That concludes our look at logical fallacies. There are certainly many others, but these are the big ones and then some, and most of the others are just subcategories of some of these. Learn these fallacies, and become handy with them. You'll find that you can easily recognize them in almost every argument someone makes, and then you're well equipped to stop them in their tracks, and require them to instead make a non-fallacious argument. Doing so strips away the bulk of the meat from the arguments of most people who advocate things that aren't evidence-based, and places you handily in a commanding position.

22. How to See Your Aura

Today we're going set up a special camera and view a mystical energy field surrounding your body that's normally visible only to certain sensitive people. Our subject today is aura photography. A listener from Chile wrote in with the following account of a late-night infomercial aired on Chilean television:

> There was a guy, by the name of Harold Moskovitz, giving a show entitled "Desarrollo Luz Dorada", or "Golden Light Development". He was claiming to provide all sorts of healing through esoteric ways such as reading ones aura and then providing techniques for resolving problems found within it. Basically the show was claiming to provide relief for pretty much any aliment, and all one has to do is attend a seminar or two, paid of course, in order to learn the mystic ways. He had numerous people swearing to have been cured of various serious ailments, and one woman even held up an x-ray of tumors she had that were now magically cured! Chile is a fine country but it does have a significant number of people that do not have access to a decent education or health care and this con man is basically telling them he can cure cancer.

Well, there's nothing new about psychics claiming to see your aura. What's missing is any kind of a half-decent explanation for what this aura supposedly consists of. It's really easy to throw around scientific sounding terms like "bioelectromagnetic fields" or "life energy", but such terms do not have any legitimate meaning. We have to ask some basic questions. Does this thing called an aura really exist? Does it convey any useful information about the person? How might it be possible that some people can see it or sense it, while others cannot?

To answer the first question, do auras actually exist, we have to abandon untestable verbal claims from psychics who say they can see them but offer no evidence, and look instead to testable evidence. Namely, aura photography. There are three types of aura photographs. The first, which is sometimes seen in video and is usually in color, is simple infrared photography. To take an infrared photograph with a conventional film camera, simply use infrared sensitive film. An

infrared photograph shows heat. A shot of a dead object at room temperature appears black or at the same color as the general background, but a living person or other warm object appears white, or in a color video, at a warmer temperature. Charged particles near the skin surface, or near the surface of any warmed object, are excited by the radiated heat and will appear as a glowing band around the person or object. Simple heat does not have any of the mystic qualities attributed to auras, and can be produced equally dramatically with any dead object if you just warm it up.

The second type of aura photograph is called Kirlian imaging. It's named for Semyon Davidovich Kirlian, the Armenian electrician who discovered it in 1937, though it didn't really become popular in the world of aura enthusiasts until the 1970's. Kirlian images of auras are all black except for the aura itself, which is manifested as a thin band of jagged white surrounding an object. To take a Kirlian image, the subject to be photographed is placed on a photographic plate which is electrically isolated above an aluminum electrode. Another electrode is connected to the subject, and the resulting image is thus burned onto the photographic plate. Kirlian described this as "bio-plasma", an image of what he called the "life energy" of the object. Scientists call this effect a corona discharge, and you get the same result using any conductive object. It does not have to be alive; a coin will work just as well. Corona discharges have been well understood since the 1700's, and they have nothing to do with "life energy" or the psychic state of an object.

Finally, we have the latest, greatest, and stupidest of aura photography techniques, which was developed in 1992 by an electrical engineer named Guy Coggins, and is called the AuraCam 6000. This produces brightly colored pictures showing the person with superimposed brightly colored clouds of light around them. To take an AuraCam picture, the subject sits and rests his hands on the leads of a galvanometer, the same device that Scientologists call an E-meter. The AuraCam takes a conventional photograph of the person which is loaded into a computer. The computer software then synthesizes an image of colorful clouds said to be based on the galvanometer measurement. This colorful halo is then superimposed onto the image of the person, then it's printed out, and presto, you have your mysterious aura photograph. According to Coggins' web site:

> Our technologies produce an electronic interpretation of what we believe the Aura would looks like. It does not

photograph the actual Aura. There's nothing that exists which can do this.

And how about those colors? How are they determined? Coggins actually gives several different answers on his web site: that the colors are determined by corresponding electrical frequencies; that the colors were chosen in consultation with psychics to produce the same colors seen by psychics; and that they are based on the writings of Dr. Max Luscher, correlating personality with color preferences. I guess you can take your pick of how you prefer to believe the colors in your picture were determined.

To Coggins' credit, his web site is very clear that these images are not suitable in any way for making medical diagnoses, and that they only represent the software's interpretations of the galvanometer reading. He is also very clear that these devices are primarily for making money taking pictures at fairs, and he provides pricing guidelines and lists of events where you might choose to take your AuraCam. I wish that all peddlers of pseudoscience were that honest.

So an examination of aura photography reveals no useful, testable evidence that auras even exist at all. If we can't establish any reasonable foundation of evidence that they might exist, it's premature to address our second question, whether any useful information can be derived from studying auras. So let's look at our third question: How might it be possible that some people can see auras?

Nobody has ever suggested a plausible mechanism for how such a thing might be possible, but if it is, there's a big fat chunk of dough sitting there waiting for anyone who can do it. James Randi has a million dollars in his vault that's yours if you can see auras. Nobody has tried since 1989, when the prize was only $10,000. On the TV show *Exploring Psychic Powers Live*, a psychic claimed that the auras she could see stretched five inches beyond the person, and that she could see the aura extend beyond the top of a screen that a person was behind. Ten screens were presented, and by prior agreement between her and Randi, she needed to score 80% correct predicting which screens had a person behind them. Random guessing would have resulted in a score of 50%, but she scored only 40%.

Can you or someone you know do better than that? If you can, Skeptoid can qualify you to apply for the million dollar prize. Just come to the Skeptoid.com web site and click on the One Million Dollar Paranormal Challenge.

Since that highly publicized failure on live television, author Robert Bruce, in his book Auric Mechanics and Theory, has taken the classic step of moving the goalposts. He states that auras cannot be seen in complete darkness or if any part of the person emitting the aura is obscured. This special pleading presumes that auras have some quality that places them on a higher plane than what might be testable using basic blinded experiments. So what it all boils down to is that the only supporting evidence for the existence of auras is the logical fallacy of special pleading that claims they are beyond the threshold of detection for anyone except self-described psychics whose claims must be allowed to be untestable.

Well, that's bogus, and it's childish. If Robert Bruce had anything in his book that could withstand any kind of scrutiny, he could easily be a million dollars richer. Until aura advocates make some kind of reasonable, quantifiable description of their phenomenon, any discussion of auras — including their photography, their meaning, or their paranormal detection — should be treated with extreme skepticism.

23. Who Kills More: Religion or Atheism?

Hide that Bible in your pocket as the guard hustles you down the snowy road on your way to eventual death in Stalin's Gulag, for today's subject is the debate over whether more people throughout history have been killed in the name of religion, or in the name of atheism.

Atheist authors like Christopher Hitchens, Michael Shermer, and Sam Harris are always debating religious authors like Dinesh D'Souza, William Dembski, and Alister McGrath about whether or not God exists, or whether or not religion is good for the world. And, as predictably as the sun rises, these debates nearly always devolve into the argument of which side is responsible for the greatest death toll throughout history. Which is a more terrible killer: religious fundamentalism, or the lack of religion?

Christians charge that the most killing in history has come from modern atheist regimes. Adolf Hitler led Germany during World War II when he executed six million Jews in the Holocaust, three million Poles, three million Russian prisoners of war, and as many as eight million others throughout Europe. Joseph Stalin was the General Secretary of the Soviet Union following the Russian Revolution until his death after World War II. Between 10 and 20 million Soviets and German prisoners of war died under his regime, depending on how many famine victims you count, from Gulags, execution, and forced resettlement. Mao Zedong, who led China for more than a quarter of a century following World War II, created the Great Leap Forward and Cultural Revolution programs which collectively killed unknown tens of millions of Chinese, most of them in public executions and violent clashes. Pol Pot led the Khmer Rouge in Cambodia during the 1970's, when as many as 2 million Cambodians, or as much as 20% of the population, died from execution, disease and starvation.

History is full of uncounted massacres by armies carrying a religious banner, though most such episodes were in ancient times with much less efficient killing technology and microscopically smaller populations. The number of religious exterminations of entire villages throughout history is innumerable, though most had body counts only in the hundreds or thousands. Alexander the Great is estimated to have executed a million. 11th century Crusades killed half a million Jews and Muslims. Genghis Khan's massacres of entire populations of cities probably totaled a million. The Aztecs once slaughtered 100,000 prisoners over four days. An unknown number, probably in the millions, died in the Devil's Wind action in Colonial India. Up to four million Hindus, Sikhs, and Muslims died in post-Colonial India. The Ottoman Empire massacred two million Armenians over the years. Franco's Spanish Civil War killed a hundred thousand. A million have died in Rwanda, half a million in Darfur. And Christian vs. Muslim violence has obviously dominated our headlines for a decade, totaling somewhere in seven figures.

So who has been the worst throughout history: secular regimes or religious regimes? Obviously the big numbers come from the 20th century superpowers (China, Russia, Germany) so the answer depends on how you classify those. And this is where the meat of these debates is usually found, splitting hairs on which regime is atheist, which is merely secular, which is non-Christian and thus fair game to be called atheist. Hitchens points out that Stalin's government had all the trappings of religion, including Orwell's totalitarian theocracy, and thus it's merely a play on words to say that it was not religious. Pol Pot was raised a Buddhist monk who grew up to execute Buddhist monks, along with anyone else he could lay his hands on. Whole books have been written on the occult underpinnings of Nazi Germany, the symbology of the Norse gods, to say nothing of the claims that Hitler was a Christian, Hitler was a Jew, and his own writings expressing the kinship he felt with the Muslims. A favorite counterpoint raised by Christian debaters is that these despots practiced Social Darwinism and were thus atheists by definition. In summary, the winner of these debates is the one who can convince the other that the big 20th century genocidal maniacs were motivated either by religion or by a desire to destroy religion. The entire debate is the logical fallacy of the excluded middle.

Here's the thing. If you write a book called *God Is Not Great: How Religion Poisons Everything*, you sell a lot of books. If you write a book called *What's So Great About Christianity* on the evils of atheism, you also sell a lot of books. If you say that neither extremist viewpoint makes any sense, you end up doing a podcast and working as a greeter at Wal-Mart directing customers to the section where they sell Hitchens and D'Souza books. The truth is less incisive, it's less inflammatory, it raises no ire, and it draws no audience.

And that truth, as I've said time and time again, is that people are people. No matter what segment of society you look at, you'll find good people and you'll find bad people. You'll even find, as has been said, that the line between good and evil cuts through every human heart. Certainly there are people in the news who kill in the name of religion, but just because they kill in the name of religion doesn't really mean they kill *because* of religion. The Islamic militants who cut off Nick Berg's head are not nice men who would have otherwise been his best friend, if it weren't for their religious convictions forcing them into this grievous act. They are base murderers, and they should be punished accordingly, I don't care whether they go to church or not. Killers don't really kill *because* of their religion. Neither does a lack of religious convictions cause one to run wild in the streets with a bloody axe and a torch. Religion is a convenient banner for many to carry, but there are plenty of other banners available as well, and if it wasn't religion, they'd do their deeds under some other justification, if they care to even have one. The real reason they do their evil deeds is that they're human. Humans are very smart, very capable; and when we want something, we generally find some way to get it, even if that means killing someone or committing genocide.

By writing this chapter, I'm going to be called an apologist for atheist genocide. My dismissal of the entire argument as pointless and fallacious will be interpreted as a dodge from advocating a weak position. So go ahead and make that charge, if you're still convinced that this is argument that can ever have a useful conclusion. I'm convinced that arguing either side is merely an opportunistic way to tingle sensitive nerves and sell a lot of books. And, I'm convinced that any discussion of the religious causes of genocide is a divisive distraction from the

more worthwhile investigation into the true cultural and psychological causes. We are human beings, and we need to understand our human motivations.

So I am no longer going to participate in the childish debate of what religion has killed more people in history, because it doesn't matter. The way I see it, you might as well debate what color underpants are worn by the largest number of killers, and try to draw a causal relationship there as well. Religion does not cause you to kill people, and it certainly doesn't prevent you from killing people. Let's stop pretending that it does either.

24. Orang Pendek: Forest Hobbit of Sumatra

Ssssshhh! We're in a dense jungle on the island of Sumatra, quietly making our way toward a brownish, three-foot-tall ape that one of our party spotted walking upright. Stop, he sees us! Wow, look at him. Note the scientifically plausible facial features and body geometry. Look at his small feet, which leave tracks surprisingly like those of the sun bear. Let's see if we can move a little closer — oh, there he goes! Watch him swing expertly up into the trees, and — wow, he's gone, just like that. Isn't it amazing that of all of us holding cameras, nobody thought to take a photograph? Well, just punch ourselves in the head for that one. Apparently, the orang pendek has some mystical quality that makes even the most dedicated of eyewitnesses forget to use their cameras.

Well, here's one convincing factoid about the orang pendek: It sounds a lot like orangutan, so it's probably a relative or subspecies, and not some ridiculous cryptid with a wild sounding name like Bigfoot or Abominable Snowman. In fact, the similarity in names is not much help at all. *Orang pendek* is simply Indonesian for "short person", just as *orang utan* means "forest person". If you were hoping that orang pendek's legitimate-sounding name meant that it has some zoological classification, you are disappointed.

Sometime in the 1980's, cryptozoologists began reading early 20th century accounts from Dutch settlers in Sumatra, and found that a few of their numerous reports about the strange animals they encountered there could be generally corroborated with one of the numerous characters from local Indonesian jungle lore, the orang pendek. Considering the large number of Dutch stories, most of which had nothing to do with any kind of ape-like creature, and the even larger number of fanciful native legends of magical forest creatures, this connection made by cryptozoologists was really quite a stretch.

But it stuck, and now orang pendek is a firm fixture in the cryptozoology files.

So much so, in fact, that in the 1990's a pair of British cryptozoologists named Debbie Martyr and Jeremy Holden began a 15-year search in Sumatra. They interviewed hundreds of natives, set up motion-triggered camera traps, made plaster footprint casts, and tramped along hundreds of miles of jungle trails. Debbie formed a detailed and specific description of orang pendek:

> *[It is] usually no more than 85 or 90cm in height — although occasionally as large as 120cm. The body is covered in a coat of dark grey or black flecked with grey hair. But it is the sheer physical power of the orang pendek that most impresses the Kerinci villagers. They speak in awe, of its broad shoulders, huge chest and upper abdomen and powerful arms. The animal is so strong, the villagers would whisper that it can uproot small trees and even break rattan vines. The legs, in comparison, are short and slim, the feet neat and small, usually turned out at an angle of up to 45 degrees. The head slopes back to a distinct crest — similar to the gorilla — and there appears to be a bony ridge above the eyes. But the mouth is small and neat, the eyes are set wide apart and the nose is distinctly humanoid. When frightened, the animal exposes its teeth — revealing oddly broad incisors and prominent, long canine teeth.*

With all of their hardware and determination, you'd think they would have gotten a photograph. But they never did. Both Debbie and Jeremy claim to have seen orang pendek on multiple occasions, but unfortunately, neither thought to employ that camera they were holding. Not even a hastily shot blurred photo of the animal running away? And yet they both saw it on several occasions? Hmmm.

More recently, two British dudes, Adam Davies and Andrew Sanderson, have been traveling around Sumatra trying to collect evidence. They brought back footprint casts and some hairs. The hairs were analyzed by microscope and determined to be from an unknown primate; and then their DNA was analyzed and found to be disappointingly human. So much for objective microscope analysis performed by cryptozoologist proponents.

There have never been any reports of orang pendek corpses or bones or body parts preserved in villages like the Tibetans do with Yeti skull caps (or goat skull caps, take your pick), so what evidence does exist of orang pendek? Well, there's nothing at all that a scientist would call evidence. There is tons of anecdotal evidence in the form of ancient legends and verbal reports, but none of that can be tested. There are footprint casts, which tend to be dismissed by most primatologists because they are indirect evidence of indirect evidence of something that's said to leave footprints exactly like those of a child. When you analyze footprints, you're up against some pretty long odds. The Bigfoot guys face this same problem. You can hold a plaster cast in your hand and measure it and say all sorts of stuff about it, but it's never good evidence. You can hold it in your hand, and yet all you know of its origin is that the person who made it gave you an untestable verbal claim that it came from a footprint-shaped hole in the ground; and even assuming that footprint-shaped hole in the ground was there, and was not made by the guy himself, it's still of unknown origin. This is why a footprint cast can never rise above the status of anecdotal evidence. But such anecdotal evidence does still have value. You can form hypotheses from it, such as, "Maybe an orang pendek does exist in the area where this cast is said to be from," and now you have a hypothesis that can be tested. We've already had a number of people out there in Sumatra testing this hypothesis, and so far they have zilch.

It seems a shame to discard all the eyewitness accounts; moreover, it seems scientifically irresponsible. These eyewitness accounts have been coming for hundreds of years. Surely all these people must be seeing something, right? Well, again, when we in the brotherhood do what we call "science" we have to sort testable evidence from untestable evidence. Untestable verbal accounts don't prove a thing, but they do give us stuff like starting points for where to search for actual testable evidence.

They clearly do have value. Debbie and Jeremy assembled a vast collection of such stories and followed every plausible lead to search for testable evidence.

So why didn't they come up with anything? Are all the eyewitness accounts and ancient stories wrong? Not necessarily. Sumatra is a big place and we're looking for a tiny little monkey walking around. What else might account for the stories? There are several reasonable possibilities. Gibbons and orangutans both live in Sumatra and could be responsible for all the sightings. Gibbons are the right size and color, but only walk on their hind legs for a short time. Young orangutans are the right size, but they generally live further to the north and orang pendek is usually reported to be a different color.

Earlier I mentioned the sun bear as a candidate for the footprints. Discounting some of the lore that says orang pendek's feet point backwards to fool trackers, the footprints are generally said to look just like those of a seven-year-old child. The sun bear, with its narrow feet and claws positioned just like toes, also leaves footprints that are said to look just like those of a (what was it again?) oh yes — a seven-year-old child.

There is a stranger possibility that has been opined by some. The 2003 discovery of *Homo floresiensis* on Flores, another Indonesian island, was widely reported in the media as a "Hobbit", a new species of early human who lived a mere 12,000 to 18,000 years ago. What we have here is actual hard evidence that a creature, roughly similar to orang pendek in size and some other respects, did live in the vicinity at one time. This doesn't suggest that it might still live in the vicinity, but the possibility is always there. It's just really unlikely, considering that it would have had to live side-by-side with humans *(Homo erectus* first came to the region an estimated two million years ago). You'd think that in all that time, there would have had to have been some crossing of paths beyond the isolated village story or two.

At least one remote population in Sumatra has a legend of small forest people that they call the ebu gogo, but modern correlations with orang pendek are really more the work of overzealous cryptozoologists than the result of any real academic historiography. Such overly optimistic correlations have drawn claims that the ebu gogo is known to have existed as recently as the year 1900, but it turns out that the source of such claims are merely stories from remote tribes of the form "My grandfather used to tell how their tribal elder got magical advice from an ebu gogo in the forest." Keep in mind that Sumatra is so diverse and fragmented that there are still 52 languages spoken there. Sumatra is a supermarket of folklore. I doubt you could find any ancient population anywhere on the island that's not going to have a dozen such legends, whether that tribe ever actually encountered *Homo floresiensis* in its ancient history or not. In such an environment, it would not be a tall order to substantiate just about any crazy story you want, just by matching it up with some local legend.

So if you travel to Sumatra and plan to spend some time camping in the jungle, I wouldn't worry too much about a tiny ape running around your campsite and wreaking havoc, or beating its little chest in a cute little Tarzan display. Maybe there is an orang pendek, but so far, if you want to believe in it, that's all you have to go on: Your own belief.

25. Medical Myths in Movies and Culture

When I first thought of this subject it sounded like a great idea, because the way TV and movies abuse our understanding of medicine and the human body has always bugged the hell out of me. But now that I've put in the research and checked out all the facts, I realize that I'm merely being a huge party pooper. If you've enjoyed believing in some of these fancies, you're probably going to be mad at me. Or better yet, just proclaim that I'm on the payroll of corporate interests, ignore everything I have to say, and go on believing that eating chocolate causes acne.

And that's as good a starting place as any. Folk wisdom tells us that eating chocolate causes acne, or that the oil from cheap greasy food like cheeseburgers or french fries will ooze right out through your skin and cause pimples. Fortunately, numerous trials have been done, and we've learned that groups eating the suspect foods don't get any worse acne than the control groups eating healthy food. So you can keep right on chowing down. The true causes of acne are heredity, hormonal changes associated with adolescence, stress, and bacteria, though some recent studies have found minor correlations with consumption of foods high in iodine. So don't drink iodine if you're pimply.

If you enjoy the taste of chocolate, you've probably also heard that taste buds are arranged on the tongue in different regions, and each region is sensitive to a particular taste. Bitterness is sensed on the tip of the tongue, sweetness on the edges, and so forth. Turns out this is another myth too. Every taste bud senses all flavors. This explains why it never works when you try to test that old story by squeezing lemon onto just little dots of your tongue, and find that it's pretty horrible no matter where you put it.

So long as we're talking about food and the senses, let's mention the old tip of improving your vision by eating carrots. As it turns out, the only connection between carrots and eyesight is the vitamin A that your body derives from the beta-carotene in carrots. You can eat all the vitamin A you want and it won't improve or otherwise affect your vision at all. If you have a severe vitamin A deficiency, it can lead to one cause of blindness. It's not quite clear how this story got started, but one source says it was a misinformation campaign by the Royal Air Force in World War II to explain the effectiveness of their night fighter pilots — the actual classified explanation being their new radar system.

Another food that's said to help is orange juice and cookies to replenish your blood sugar after you give blood. The problem with this is that there's no reason your blood sugar would be any lower or require replenishment after donating. Giving blood just sucks a safe amount of your blood reserve out of your body; it does not change or weaken the blood that remains. Nor is the rest of your body dehydrated after giving blood, so there's no more reason that you would need to rehydrate than there might otherwise be. Some people might get nervous or faint from the process, and the refreshment might help to relax them; but there is no medical need for juice, cookies, flowers, kind thoughts, or anything else.

While we're on the subject of blood, let's talk about one of my pet peeves from Hollywood. In Pulp Fiction, John Travolta stabbed Uma Thurman with a syringe full of epinephrine directly into her heart to cure a drug overdose. In The Rock, Nicolas Cage does the same thing to himself to counter the effects of poison gas. Wow, makes for a dramatic movie scene, doesn't it? And now, since it was such an exciting scene, practically every TV and movie writer thinks it's real and puts it into about every other show. Well I'm sorry to burst your epicardium, but according to emergency room doctors, there is no actual medical treatment that involves the dramatic stabbing of a huge needle directly into the heart — certainly not through the breastbone or in any kind of violent or forceful manner. The way to get any medication into the heart is to simply inject it into a vein. No driving musical soundtrack required.

Of course, if you did accidentally kill your friend by stabbing them in the heart with a syringe, you might get to test the old story that their hair and fingernails will continue growing after death. There's no truth to this either. Metabolism stops at death, so there is no possible mechanism by which new hair or fingernails could be created. This rumor probably got started because a dead body's soft tissues dry out and shrink and pull away, exposing more of the hair and nails. Egyptians didn't really look that creeped out in real life.

Neither does hair grow any thicker or darker after it's been cut. Everyone's been told this, but nobody seems to believe it. Hair is made of dead cells. There is no metabolism or living nerves in hair, thus no mechanism by which the tip of a hair follicle could communicate that it had been cut back to the root to stimulate additional growth or the development of new hair follicles. Some people have longer hair and some people have shorter hair, both on their head and on their bodies, and the speed of growth and the lifecycle of the follicles is determined by your genes. It can't be changed, certainly not by anything as simple as cutting. A cut-off tip of hair is more visible than the finely tapered natural end, which probably explains why so many people still believe this; but that cut hair will never be as long as the natural hair.

Here's a good one, and it's a personal favorite because it happened to a friend of mine. He staggered up out of the water in Cancún with a Portuguese Man-o'-War stuck to his shoulder, tentacles glued all over his torso, and collapsed on the beach writhing in agony. While someone called for help, some gringo SCUBA divers on hand offered to help the way they knew best: All six of them unzipped and hosed him down liberally. Sadly for my friend, the old home remedy of urinating on a jellyfish sting only makes things worse. While vinegar will in fact block any remaining stinger cells from firing, urine contains ammonia, which causes the stinger cells to fire. Let's just say it was a bad day for my friend.

On a less painful subject, who among us does not have a mom who has whacked us with a ruler for cracking our knuckles? Folk wisdom says knuckle cracking leads to arthritis or joint enlargement. As a lifelong knuckle cracker, I can confidently attest to no ill effects. Nor should I expect any. The crack you hear is simply the popping of bubbles within the synovial fluid as the ligament is stretched, as hard to believe as that sounds. It causes no problems and has no cumulative effects. While the popping itself is harmless, the repeated stretching of the ligaments can lead to, well, stretched ligaments, but this too is unrelated to either arthritis or enlargement. Never pay attention to your mother. I'm pretty sure my own mom's not reading this, so I can say that.

Have we had enough of these yet? How about one more?

Remember in Beverly Hills Cop, when Judge Reinhold said to Sgt. Taggart that the average person has five pounds of undigested red meat in their bowels? Urban myths like this one are largely responsible for the popularity of colon cleansing in holistic medicine. The hose enters through the exit and some solution is pumped in to rinse out the daily output of your digestive system. This is based on a wholly erroneous assumption, that you have old junk or "toxins" built up in there. You don't. The digestive system is an active, working, one-way conveyor belt. Nothing stops and stays in there. If it does, that's called a blockage, which is a serious medical emergency. Unless you are in the emergency room right now with an intestinal blockage, there is nothing in your intestines older than about 24 hours, other than bacteria which live safely tucked away in the walls. If you've ever had a colonoscopy, you know that all you have to do is stop eating and drink water for a day or two and your bowels will be as clear as a Smurf's conscience.

Whenever you hear a story or a rumor about the human body that seems unusual or doesn't make any obvious sense, be skeptical. It may be true after all — the body is a fascinating machine that's full of surprises — but it's always best for your health to follow a skeptical process and determine the facts for certain.

26. Aliens in Roswell

Hang onto your tinfoil helmet, because today we're going to rocket into the history books and see for ourselves exactly what fell out of the sky in Roswell, New Mexico in 1947.

In July of that year, a balloon train came down on the Foster Ranch, 75 miles northwest of Roswell, New Mexico. Rancher "Mac" Brazel, who had been reading about flying saucers, reported it to the local Sheriff, who in turn reported a crashed flying saucer to a Major Jesse Marcel at Roswell Army Air Field, but not before the local press heard about it. The debris, totaling some five pounds of foil and aluminum and described in detail by Mac Brazel, was recovered by officials from Roswell Army Air Field. These balloon trains were long ultra low frequency antennas designed to detect Soviet nuclear tests, held aloft by a large number of balloons, and were known as Project Mogul. With Marcel's press release in hand, the *Roswell Daily Record* reported that a Flying Saucer was captured, and the following day, printed a correction that it was merely a weather balloon, along with an interview with Mac Brazel, who deeply regretted all the unwanted publicity generated by his misidentification.

It should be stressed that this was the end of the incident, and nothing further was said or done by anyone, until 1978 (that's 31 years in which nobody remembered or said anything), when the *National Enquirer,* on what must have been a slow news day, reported the original uncorrected news article from the *Roswell Daily Record.* UFO fans went nuts. Stanton Friedman, an obsessed UFO wacko, started interviewing everyone he could find who was still alive who had been connected with the incident and began constructing all sorts of elaborate conspiracies. These primarily centered around Major Marcel, who agreed that Friedman's assertion was possible — that the government was covering up an actual alien spacecraft.

Two years later in 1980, UFO proponents William Moore and Charles Berlitz published *The Roswell Incident*. There wasn't much new information in this book, it was essentially a collection of suppositions and interviews with a few people who were still alive, or their relatives. Even so, by this point, it's important to note that the story really had not grown beyond the question of what debris had actually been recovered from the Foster Ranch in 1947.

Upon the book's publication, the *National Enquirer* tracked down Marcel and published their own interview with him. This was all well and good, but since there still wasn't any new information or any evidence that Roswell was anything other than the Project Mogul balloon, things quieted down for a long time.

The story finally started to break open for real in 1989. The TV show *Unsolved Mysteries* devoted an episode to an imaginative "reconstruction" of what some of these authors had written. The national exposure of a TV show reached a man named Glenn Dennis, who was quite elderly by now but who had worked as a young mortician in Roswell in 1947, and had provided contract mortuary services to Roswell Army Air Field. Dennis contacted Stanton Friedman, and told him the story that was to become the basis for almost all modern UFO lore.

Virtually all popular details of the story of an alien crash at Roswell are based upon the personal recollections of Glenn Dennis. He hadn't thought about the subject for 42 years, until he saw the TV show. Suddenly he started putting two and two together, tying together bits and pieces of this and that from his memory, and with the help of Stanton Friedman, connected the dots and wove the fabric of modern Roswell mythology. Authors Schmitt and Randle published Friedman's interview in *UFO Crash at Roswell*, published in 1991. This was the point that all the best-known details were invented: the multiple crash sites, the alien bodies recovered, the child-sized coffins, aliens walking around the base, a red-haired colonel making death threats, and the disappearance of a nurse who knew too much. 1991, sports fans; not 1947.

What's worse is that Glenn Dennis' memory doesn't seem to handle dates very well. Air Force researchers have successfully been able to corroborate nearly all of Dennis' recollections, but what they found was that while nearly all the events he remembered did in fact happen, they happened over a span of 12 years; not around a single incident in 1947. Let us now go through a few of the most significant points from the pop history of the Roswell incident, one by one:

1. At his mortuary in Roswell, Dennis received a call asking him to give a ride to an airman injured in a traffic accident.
2. At the base, Dennis noticed an ambulance filled with wreckage that looked like the bottom of a blue canoe, guarded by MP's.
3. Dennis was ejected from the base by a big red-headed colonel who threatened him with death if he revealed anything of what he had seen.

4. Dennis tried to telephone a nurse that he knew, but was rebuffed by a head nurse nicknamed Slatts, and whose real name was Captain Wilson.
5. The next day, Dennis successfully met with his nurse friend, who was quite upset over an autopsy on three bodies which she described as black, mangled, and little.
6. Dennis learned of the inability of staff to perform an autopsy due to overpowering fumes from the bodies, which were then hastily moved.
7. All further efforts to reach the nurse were unsuccessful, he was told she was deceased, and Dennis never saw her again.
8. A creature with a huge head was seen walking into the base medical center "under its own power".

And now let's look at what each of these events really was, and why they couldn't possibly have all been part of something that happened in July 1947:

1. Dennis' trip to the base to deliver the injured airman is not likely to have happened in July of 1947, as the rank of airman did not exist until the United States Army Air Forces became the United States Air Force in September of 1947. Note that Roswell Army Air Field was renamed Walker Air Force Base at that time as well.
2. Dennis' description of the contents of an Air Force ambulance are consistent with its normal appearance, which includes two steel panels shaped like the bottom of a canoe and painted Air Force blue.
3. Only one tall captain or colonel with red hair was ever assigned to the base, and that was Col. Lee Ferrell, who began his service at the base in 1956. Dennis also remembered that the threatening colonel was accompanied by a black sergeant, which is also virtually impossible for 1947. The Air Force did not begin racial integration until 1949. Dennis was probably recalling some episode that happened in or after 1956, and keep

in mind he was still stretching his memory over 30 years.

4. The head nurse, Captain Wilson, has been identified as Idabelle Miller, who later married and became Major Wilson. She did not begin her service there until 1956. "Slatts" was well known to be Lt. Col. Lucille Slattery, a different person, who also did not arrive at Roswell until after the Roswell Incident. In her case the timing was much closer, only one month afterward; but one month too late is still too late.

5. Dennis' nurse friend has been positively identified as Lt. Eileen Mae Fanton, who was in service there during the Roswell Incident. More on her in a moment.

6. In 1956, a KC-97G aircraft crashed, killing all 11 crewmen in an intense cabin fire. Most were missing limbs and were largely burned away, resulting in small, black, mangled corpses. Due to intense fumes from the fuel soaking the bodies, the procedure had to be hastily moved to a refrigerated unit at the commissary. Three of the bodies were autopsied at Dennis' mortuary in Roswell, which Dennis described as small, black, and mangled. It's entirely likely that Dennis' encounter with the angry redheaded colonel took place during this exceptionally emotional and difficult time for the small local Air Force community.

7. The reason Dennis ceased being able to reach Lt. Fanton was that she had been taken to Brooke General Hospital in Texas for emergency treatment of a pre-diagnosed medical condition which ultimately led to her medical retirement in 1955. This information was withheld from Dennis simply due to patient privacy laws.

8. The sighting of a creature with a huge head walking into the hospital is consistent with the injury of Captain Dan Fulgham in 1959, who was struck on the head by a balloon gondola. His forehead and face developed an extensive hematoma which swelled to quite a magnificent size. He said he didn't feel too bad, and

actually hung out smoking a cigarette and walking around with a head the size of a beachball.

Now, obviously I have to leave a lot of details out since this is a book of short chapters and can't possibly address the billions of data points that the UFO proponents have thrown out there (if you remember the chapter on logical fallacies, you'll recognize that technique as Proof by Verbosity). If you're truly interested in the actual explanation of some detail you've heard, download the *Roswell Incident* PDF report from the Air Force at af.mil, or get a copy of the free *Roswell Report: Case Closed* book by Captain James MacAndrew and published by the Air Force. These publications are not a desperate coverup by the Air Force, they were required by the General Accounting Office's official inquiry made to address the clamor of Freedom of Information Act demands from UFO proponents. They contain a tremendous amount of detail and are quite entertaining reading for anyone interested in Roswell or the Air Force's early history.

But about every year or two, someone else pops up with some new claim about what happened in 1947. Just this past year, a Lieutenant Walter Haut died, leaving a written story about having been shown a crashed alien spacecraft, but he lacks credibility. The stories he told about his experience at Roswell grew and grew over the years. And, as the president of the International UFO Museum, he had a clear commercial interest in promoting these stories. Even less believable are the stories of Philip Corso, who co-authored a fictionalized retelling titled *The Day After Roswell*. Among Corso's claims are that lasers, Kevlar, fiber optics, and integrated circuits all came from the Roswell spacecraft. Since the true origins of all of these technologies are well established in the real world, even other UFO researchers discount most of Corso's fancy tales.

So when you look at a story like Roswell, look at it skeptically. One account is of mundane everyday activities, that are fully supported by hard evidence at every step; and the other version is wild, far out, and comes in many conflicting versions, none of which have any supporting evidence whatsoever. What does a responsibly skeptical process support?

27. Death in Your Kitchen: Microwave Ovens

Today we're going to walk on the wild side and eat some microwaved food. You've never really lived until you've lived dangerously, so let's put our lives on the line by testing the claims that microwaved food and water are toxic.

First, a little background info. Microwave ovens work by passing microwave-band electromagnetic radiation over the food at 2.5 GHz. Molecules that are electric dipoles, of which water is the most efficient, rotate back and forth in this field. The friction between them creates heat. This is called dielectric heating. More complex molecules, which are not as clearly dipolar, are not affected. It's an efficient and clean way to heat food.

I first learned about the claims of danger from a chain email sent by a friend of mine who tends to believe anything that's anti-establishment or on the fringe. A few Internet searches of some keywords reveal a huge number of holistic, organic, and other alternative web sites repeating these same claims. Just to give you a flavor of how far-out these stories are, take a look at this list of *Ten Reasons to Throw Out Your Microwave Oven* widely reproduced all over the Internet. Notice that not one of them makes a specific or testable claim; they are all merely scary sentences constructed using scientific sounding words. And, as you can tell from the brief description of how microwaves work, few of these have any remote connection to fact:

1. Continually eating food processed from a microwave oven causes long term — permanent — brain damage by "shorting out" electrical impulses in the brain, de-polarizing or de-magnetizing the brain tissue.
2. The human body cannot metabolize the unknown byproducts created in microwaved food.

3. Male and female hormone production is shut down and/or altered by continually eating microwaved foods.
4. The effects of microwaved food by-products are permanent within the human body.
5. Minerals, vitamins, and nutrients of all microwaved food is reduced or altered so that the human body gets little or no benefit, or the human body absorbs altered compounds that cannot be broken down.
6. The minerals in vegetables are altered into cancerous free radicals when cooked in microwave ovens.
7. Microwaved foods cause stomach and intestinal cancer tumors. This may explain the rapidly increased rate of colon cancer in America.
8. The prolonged eating of microwaved foods causes cancerous cells to increase in human blood.
9. Continual ingestion of microwaved food causes immune system deficiencies through lymph gland and blood serum alterations.
10. Eating microwaved food causes loss of memory, concentration, emotional instability, and a decrease of intelligence.

One clue that might encourage you to regard these claims with some skepticism is that fact that ever since microwave ovens came on the market in 1954, not one person has ever exhibited a single symptom of any illness resulting from having eaten microwaved food, or from having used water that had been microwaved. Burns are the exception, but burns are caused by heat from any source; that's not unique to microwaves. But if you believe the claims by the anti-microwave fringe, whom I call the Microwave Militia, practically everyone on the planet should be gravely ill with cancer, radiation poisoning, malnutrition, and mental retardation.

The same chain email, and many of these web sites, also state that giving a plant water that has been microwaved will kill it. There is even a series of unsourced photographs of two plants, one of which withers and dies while its sibling flourishes. The awesome web site Snopes.com tested this particular claim. They took three plants of each of several types, and watered one with tap water, one with water that had been boiled over a stove, and the third with water that had been boiled in a microwave. Unlike whoever took the pictures that often accompany the chain email, Snopes actually controlled for

other variables. I'm sure you won't have to stretch your imagination very far to guess how the plants did. They all did exactly the same. Snopes has complete details and photographs on their web site. Somehow these plants managed to escape the guaranteed death sentence that believers say microwaved water carries.

This whole paranoid suggestion is based on the presumption that a microwave oven somehow changes or poisons water. If true, wouldn't you be able to perform some kind of a test on water, and see if it has ever been microwaved? Water is H_2O, whether it's ever been microwaved or not. But here's an even deal for you. If you truly believe that H_2O carries some permanent damage as a result of being microwaved, and that it's possible to detect this damage through any means you choose, there's a million dollars in it for you. As you may know, the Skeptoid podcast is a qualifying media outlet for the James Randi Educational Foundation's *Million Dollar Paranormal Challenge*. I'll walk it through for you and I'll become your biggest cheerleader. Are microwaves really a danger to humanity? If so, it would be immoral for you to do anything else but take that million dollars and use it to educate and save the world.

Probably the most flagrant error that the Microwave Militia propagates is that microwaved food or water contains what they call "radiolytic compounds" — new chemicals created by the tearing apart of molecules in a microwave. These new chemicals are said to be dangerous, cancerous, radioactive, unnatural, or otherwise harmful. This is a demonstrably false claim. Radiolysis, which is a real process and which the Militia believes creates these radiolytic compounds, is the process by which molecules are dissociated under ionizing radiation. Water can be dissociated under ionizing alpha particle bombardment, which is a natural process. Microwave radiation, as mentioned earlier, is not ionizing radiation. It is thus scientifically incapable of causing radiolysis. The differences between microwave radiation and alpha radiation are huge. With the claim that microwaves cause dissociation of water molecules, the Microwave Militia is either deliberately lying, or they are grossly ignorant of the very subject on which they claim superior expertise.

Swiss vegetarianism advocate Dr. Hans Hertel is perhaps the most vocal of the Microwave Militia fringe group. He is quoted in virtually every book written on alternative foods or holistic health. A top-selling book on Amazon called *Perfect Balance*, by an author known simply as "Atreya", writes:

> In spite of the political pressure, Hertel has continued his studies and won the support of many other scientists in Europe for his findings and methodology. Hertel concludes that microwaved food alters the blood chemistry of people

who eat it. *The manufacturing companies are trying to keep this information suppressed through court orders.*

Dr. Hertel seems to have managed to gain this claimed following even without producing the most basic of information that prospective groupies should request: a specific, testable claim about what this change in blood chemistry might be; or a single victim. He is best known for his most publicized test. In 1989, he and seven fellow vegetarians confined themselves to a hotel and consumed only milk and vegetables, prepared in different ways, for two months. When he emerged, he announced his results: That microwave ovens cause cancer and degenerative diseases, despite no cases of cancer or illness among he or his group. His research, if you want to call it that, was never peer reviewed or published in any reputable journal, and yet it has become the foundational magnum opus of the anti-microwave agenda.

You'll also find that there are a large number of studies out finding changes to the nutritional content of food that has been microwaved, and the Microwave Militia loves to point to these. Chemical reactions happen whenever any food is cooked, so this has more to do with cooking than with the cooking method. Moreover, such changes are generally well below any perceptible threshold, and have always been found to be safe.

The Microwave Militia also makes claims such as microwave ovens are illegal in Russia or other parts of Europe. This is just a straight-up lie. Microwaves are perfectly legal in Russia and everywhere else in Europe. In fact I was not able to find a single country in the world that bans microwave ovens. They're regulated, of course, like all electric appliances; but regulation should not be mischaracterized as a ban.

So what's the sum total of our evidence? Billions of people have been eating microwaved food for decades, with no ill effects, and no plausible expectation of ill effects. The best evidence put forward by anti-microwave activists is based on shameless lies and irresponsibly bad science. Thus, a truly skeptical process leads us to the conclusion that there's nothing at all wrong with microwaving your food. However, I'm drinking coffee right now made from microwaved water, and it's entirely possible that this has caused profound mental aberration, and made me spout nonsense.

28. Ghost Hunting Tools of the Trade

Television shows about ghost hunting have been popular for over 50 years, and though the basic concept is the same, recent decades have seen the hunt become less about psychics and séances and more about electronic detection gear. Just about every TV show about ghost hunting sends a crew of investigators into a building, armed to the teeth with all sorts of equipment.

The use of any kind of measuring equipment to detect ghosts is fundamentally, and completely, bogus. How can I make a blanket statement like that? Measuring equipment detects what it is designed to detect, whether that's light, heat, electromagnetism, or whatever. Thus it will only detect things that emit measurable amounts of those energies. For us as viewers to accept that some piece of handheld measuring equipment has a useful function in detecting a ghost, we must base our acceptance on the premise that ghosts are known to emit those types of energies in measurable amounts. If there were any truth to this, science would have discovered it long ago. Hospital operating rooms would have ghost detection equipment built in. Mortuaries and crematoriums would have ghost detection equipment at the top of their list. Search and rescue crews would use ghost detection equipment. If ghosts did exist and were detectable, you can bet that there would be huge industries behind it. I can't think of anything that would attract more venture capital dollars from Silicon Valley. However, no rigorous research has ever shown that ghosts can be reliably detected with hardware. It's easy to disbelieve me, but it's much harder to disbelieve the lack of interest from greedy corporate America.

So now let's look at the popular tools of the trade of ghost hunting. The important takeaway is to understand what these devices are actually detecting when the ghost hunters point them around the room, and why their crazily jumping needles and indicators are perfectly consistent with, and explained by, the absence of ghosts.

Infrared thermometers are the most blatantly misused of the ghost hunting tools, so are a great place to start. These handheld devices measure the temperature of the object they are pointed at. They work exactly like your vision, except that they are sensitive to far infrared instead of the visible spectrum. They measure surface temperatures, just like your vision measures surface colors. If you can see something, an IR thermometer can measure its temperature. However, ghost hunters use these devices to detect what they believe are cold spots in rooms. IR thermometers are not capable of detecting something without a visible surface. In fact, an IR thermometer is even less likely than your vision to see a hazy apparition. Firefighters use infrared because the longer wavelengths of infrared penetrate smoke more effectively than the shorter wavelength of visible light; so if there were a hazy invisible apparition floating in the middle of the room, infrared is perhaps the worst technology to detect it. Variations of IR readings inside a room are merely showing temperature gradations on the walls, caused by heating and AC, insulation variances, studs, wiring, or pipes behind the wall, radiant heat, recent proximity of another ghost hunter, sunlight, temperatures in adjacent rooms, or countless other causes.

Infrared motion detectors work on the same principle. If the amount of IR radiation striking the sensor changes, an alarm can be activated. Such a change is caused by a sudden change in temperature within the detector's field of view, or a significant movement by an object with a visible IR signature. A ghostly cold spot moving within a room could not be detected, unless it also cooled the walls or floors enough to trip the activation threshold.

Particle detectors are devices that measure ionizing radiation. The most common particle detectors are Geiger counters, also called halogen counters. These work by measuring cascade effects caused by incoming particles that strike molecules of halogen gas inside the detection chamber.

Typically, alpha, beta, and gamma particles are detected by these. It's not the most common of ghost hunting tools, but occasionally you will see someone pointing a Geiger counter around the room, though you may hear them describe it by any of several fancier and more high-tech sounding names. It's a Geiger counter. For a ghost to emit ionizing radiation, it would have to be an awfully sick ghost; or be composed largely of unstable radioactive metals. Ionizing particles don't just appear out of thin air, they are emitted by the decay of unstable isotopes that are typically heavy and have significant mass.

EMF meters are perhaps the favorite tools. EMF meters detect changing or moving electromagnetic fields, and are used in ghost hunting on the premise that ghosts emit electromagnetism, though this claim is rarely supported by any suggestion of what the power source might be. There are many different types of EMF meters. More affordable units, such as those typically used by television performers, need to be held precisely for a period of time at each of the three axis to get a reading, and so they are clearly not used on television in a manner that would produce any useful result. When they are, or when a more expensive three-axis meter is used, they are designed to detect the operation of electrical appliances or wiring. Ghost hunters are usually thoroughly accessorized with every electronic gizmo under the sun: radios, cell phones, flashlights, cameras, TV cameras, and other ghost hunting accessories; and all of these will produce a result on the EMF meter. Building wiring or appliances will also be detected. But, even in an environment with no electrical devices at all, the presence of the TV camera alone renders the EMF readings totally useless. Even without ghost hunting equipment, electrical wiring, or a TV camera, a sensitive meter can even detect the oscillation of a steel filing cabinet vibrated imperceptibly by footsteps. In the midst of all the absurd amounts of EMF pollution on a TV ghost hunting set, the pretense that the alleged EMF field of a ghost (who's not carrying any batteries) can be identified, is foolish.

Ion detectors are interesting animals. The few commercially available ion detectors are available online almost exclusively through ghost hunting and alternative wellness web sites, which gives some clue of how useful they actually are among legitimate science. Ions occur naturally in the atmosphere from a variety of sources: solar radiation and weather being the main ones. Also, if you go to a part of the country where radon gas is an issue, an ion detector taken into the basement can go crazy sensing airborne ions created by radon decay. Ghost hunters prefer to regard this reading as indicative of the presence of a ghost. Ion detectors can also sense the presence of static electricity, so if your ghost is carrying a large static charge, you ought to be able to see it scuffing its shoes across the carpet.

Cameras of different types are used by ghost hunters. Sometimes they'll take a conventional visible spectrum camera and snap away, in the hope that spirit orbs or other manifestations will appear on the processed film. Since this phenomenon has already been thoroughly discussed in our chapter on orbs in the previous volume, there's no need to repeat it here. Suffice it to say that all such images are well established artifacts of photography and of cameras, and well understood by knowledgeable photographers. They happen every day in photographs that have nothing to do with ghosts. Near infrared photography is the monochromatic "night shot" video that you see all the time, and that your home video camera probably offers. The light source is an infrared bulb on the camera, similar to the invisible light source inside your TV remote. These cameras record only what near infrared light is reflected from the subject, and of course they also record other near infrared sources, which are relatively common. Far infrared photography is the thermal imaging discussed previously. It's simply a visual display of the same heat sources detected by IR thermometers and motion detectors.

Dowsing rods are probably the least controversial of ghost hunting tools, in that increasingly few people accept that they have any useful function. Yet ghost hunters still employ them. And why not? A self-described psychic's untestable verbal reports are under the psychic's complete control. They cannot be tested, measured, or duplicated by others — they say only exactly whatever the psychic wants to say. Dowsing rods simply give the dowsers another way to communicate whatever they choose to communicate. Since the rods are held in the dowser's own hand, they move only when the dowser wants them to move, and do not move when the dowser doesn't want them to. No form of dowsing has ever passed any type of controlled test, and no dowser has ever proposed any plausible hypothesis suggesting that dowsing might be an actual phenomenon. It is among the most childish of pretended ghost detection methods. The only thing you can learn from dowsing is which way the dowser wants to swing his dowsing rods.

Audio recording gear is used when the ghost hunter hopes that EVP. or electronic voice phenomena, will appear on the recording. EVP's are discussed often enough to warrant their own chapter. An EVP is said to be the voice of a ghost, and the claim is that ghosts can talk perfectly well but can only be heard on an electronic recording. This means that recording gear has the ability to convert inaudible frequencies into audible ones. Engineers do not design this capability into most recording gear, since a change of frequency of perhaps tens of thousands of hertz would render all recordings completely useless and horrible to listen to. So, like they tend to do with all the electronic gear they carry, ghost hunters completely misunderstand, misuse, and mischaracterize the functions of these instruments.

When you turn on the television and you watch people pointing their gizmos around the room, acting all dramatic and pretending to detect ghosts all around them, any intelligent adult should laugh out loud. Or better yet, change the channel. Of course an intelligent adult should be free to watch whatever they want, and that's fine — but one place I will draw the line is the point where you let your children watch one of those shows and allow them to accept the silly claims as fact. Watch it and enjoy it as entertainment, if you find those people truly engaging and clever enough to be entertaining; but please, explain to your kids the science behind what they're seeing. Or, as the case may be, the lack of science behind it.

29. What Do Creationists Really Believe?

If you've listened to the news at all within the last few years, you know that there's one topic which is always in the headlines. It's more lasting than terrorism, more pervasive than politics, and more personal than global warming. It's the war over religion; specifically, having religion taught as scientific fact. Replacing science with creationism. Whether it's Tennessee v. Scopes in 1925 or Kitzmiller v. Dover in 2005, religion versus science is always front and center.

Watching the news you've seen the $27 million Creation Museum in Kentucky, the largest and newest of the several museums throughout the United States depicting Biblical literalism as an alternative view of natural history. Dioramas show early farmers using small dinosaurs as beasts of burden. Dramatic displays show how Noah's flood created the Grand Canyon and all major geological features in a few days a few thousand years ago, and even give insight into how Noah kept all the dinosaur species on board his 600-foot ark. Most reasonable people are shocked by these flagrant attacks against intelligence. Does this mean that everyone who calls himself a creationist is certifiably insane?

As we see in so many aspects of our culture, it's usually the loudest and most outrageous fringe minority that makes the most noise and gets the most headlines. Rest assured that most creationists do not believe that Jesus rode around on a saddled Triceratops. There are, in fact, a number of different types of creationism. These variations conflict with one another and are mutually exclusive, and they are at varying odds with science.

The movement called Intelligent Design is not a type of creationism, or indeed any particular set of beliefs, so it will not be included in this discussion of the various types of creationism. Intelligent Design is a blanket concept intended to show that the scientific method alone is not adequate to explain the natural world, and that a divine creator is a required component for any complete explanation of nature. All types of creationists rally under the banner of Intelligent Design with the explicit goal of getting a foot in the door to force their particular belief system to be taught as fact in public schools.

These types of creationists fall into two main classifications: Young Earth Creationists, who believe that the Earth is between six and ten thousand years old; and Old Earth Creationists, who generally accept the scientific measurement of the Earth's age at 4.5 billion years old. Within these classifications are other irreconcilable differences, which we'll now go through one by one.

Let's start with the forms of Old Earth Creationism. I'm going to describe five basic types. Philosophers and adherents will probably quarrel with my chosen five, as there are others, and there are undoubted overlaps between these, and many believers combine aspects from two or more. But let's stick with these five as being representative. Here they are, in order of how well they reconcile with science, starting with the best:

1. **Theistic Evolution.** This is the Catholic Pope's officially stated position, and it's embraced by many real scientists of faith. Theistic evolution accepts both the geologic and biologic records, including modern evolutionary synthesis, and posits that these are simply the tools God chose to create the natural world. Theistic evolution allows and embraces scientific research and permits the acceptance of new information.

2. **Evolutionary Creationism** also accepts the geologic and biologic records, and makes its creationist distinction in that there were a literal Adam and Eve who were simply the first spiritually aware humans, though they came into being in the same way as all early humans.

3. **Progressive Creationism** goes one step farther. Progressive Creationism accepts the geologic record, and much of the prehistoric biologic record, including the true age of dinosaurs and other early lifeforms, but

believes that the creation of humans and perhaps other modern animals was a special creationism event as literally depicted in Genesis. Thus, there can be no biological link between humans and early hominids from the fossil record.

4. **Day-Age Creationism** is the belief that the six days of creation were really six geological epochs. Usually some effort is made to reconcile specific days in Genesis to specific epochs in Earth history, but since things didn't really all happen separately and consecutively like in Genesis, such efforts are generally somewhat ham-handed. But at least they're trying. Day-Age Creationism is what Jehovah's Witnesses advocate in their Watchtower pamphlets.

5. **Gap Creationism** is about as far as the Old Earth model can be stretched. This model attempts to unify the true age of the Earth as measured by science with the literal Biblical account. Jimmy Swaggart advocates this model. Gap Creationism states that the first verse of the Bible, God created the heavens and the Earth, was followed by a "gap" of 4.5 billion years, during which time not much happened. And then, the literal creation of Genesis took place in six days about six to ten thousand years ago. Necessarily, this model has to abandon evolution completely, although it adheres to proper geology.

Now we move to the other half of creationist models, the Young Earth Creationism. Here we are forced to completely abandon reason and rationality. There are really only two main camps, and as you can see, they are completely at odds with one another, agreeing only on a single point: That the Earth did not exist ten thousand years ago. Let's now examine these two types of Young Earth Creationism, and once again we'll take them in order of how closely they adhere to real science:

1. **Omphalism.** This is named after the 1857 book *Omphalos*, published two years before Darwin's *Origin of*

Species, which explained that the fossil record was God's way of making the Earth appear to be old. Omphalos is Greek for navel, and the Omphalists believe that Adam and Eve were created with navels, thus having the appearance of being created through normal evolutionary biology. Adherents to Omphalism fully accept every scientific measurement of the age of the Earth and every discovery of modern biology, with the important exception that all such discoveries are wrong: God only wanted to make us think that the Earth is 4.5 billion years old, and that life evolved from lower forms. A true scientist doing real research can be an Omphalist. He will arrive at the correct conclusions, though he will believe that his measurement is merely what God wants him to see.

2. **Modern Young Earth Fundamentalism.** Here is where the train jumps completely off the tracks. Modern Young Earthers, for lack of a better name, are the ones behind the Creation Museum discussed earlier. They honestly believe in alternate versions of virtually every science known, throwing away every shred of modern science that doesn't point to the age of the Earth as 6,000 years. They literally believe in Adam and Eve (without navels) and all the dinosaurs on Day 1, fossilization taking only a few hundred years, and all major geologic features having been created in a few days in Noah's Flood. They reject evolution, cosmology, geology, and every science that supports them; which, by extension, eventually includes every scientific discipline. However, in their minds, they don't reject them at all; they fully embrace completely wrong, misinterpreted, misunderstood, and misrepresented versions of them. Their worldview is based absolutely on the Bible as a perfect, unerring, literal historical account. As a followup, they have invented their own versions of natural sciences that they pretend supports this view. It is not possible to be a thoroughly researched Young Earther and still retain any grasp on rationality. This is

the group making the overwhelming majority of noise in the media and modern culture, but it's not clear how large of a group this really is. They have the largest and loudest web presence, with AnswersInGenesis.org and the Discovery Institute, though out of 3.2 million Ph.D.'s worldwide they've only been able to find 700 who agree with their science, according to their list maintained at DissentFromDarwin.org. This represents 2% of 1% of people with advanced academic degrees.

So as you can see, the battle is not simply between science and creation. It's really more between the various forms of creationism, and especially between the modern Young Earthers and everyone else. There are perfectly rational ways to blend what we've learned through the scientific method with divine guidance, if that's your cup of tea. There are even reasonably, or at least relatively, rational ways to accept the gist of Genesis and still maintain a grip on reality. The majority of creationists are not entirely disconnected from reason. Even people like the Jehovah's Witnesses, who are often thought of as fringe fundamentalists, keep their beliefs reconciled with modern science. So long as this focus is maintained, we can be reasonably assured that our educational system is not headed for the proverbial rubber room.

Old Earth Creationism	Accept Geology	Accept Biology
Theistic Evolution: Evolution by natural processes is the tool God used	Yes	Yes
Evolutionary Creationism: Adam and Eve were the first spiritually aware humans	Yes	Yes
Progressive Creationism: Humans were a special creation event	Yes	Most
Day-Age Creationism: Six days of creation were six geological epochs	Yes	Some
Gap Creationism: 4.5 billion year gap between Genesis 1:1 and 1:2	Yes	Some
Young Earth Creationism	Accept Geology	Accept Biology
Omphalism: Earth was created with the appearance of age and of evolution	Yes	Yes
Young Earth Fundamentalism: Invented versions of all natural sciences to explain Earth's age as 6,000 years	No	No

30. The Detoxification Myth

Today we're going to head into the bathroom and suck the toxins out of our bodies through our feet and through our bowels, and achieve a wonderful sense of wellness that medical science just hasn't caught onto yet. Today's topic is the myth of detoxification, as offered for sale by alternative practitioners and herbalists everywhere.

To better understand this phenomenon, it's necessary to define what they mean by toxins. Are they bacteria? Chemical pollutants? Trans fats? Heavy metals? To avoid being tested, they leave this pretty vague. Actual medical treatments will tell you exactly what they do and how they do it. Alternative detoxification therapies don't do either one. They pretty much leave it up to the imagination of the patient to invent their own toxins. Most people who seek alternative therapy believe themselves to be afflicted by some kind of self-diagnosed poison; be it industrial chemicals, McDonald's cheeseburgers, or fluoridated water. If the marketers leave their claims vague, a broader spectrum of patients will believe that the product will help them. And, of course, the word "toxin" is sufficiently scientific-sounding that it's convincing enough by itself to many people.

Let's assume that you work in a mine or a chemical plant and had some vocational accident, and fear that you might have heavy metal poisoning. What should you do? Any responsible person will go to a medical doctor for a blood test to find out for certain whether they have such poisoning. A person who avoids this step, because they prefer not to hear that the doctor can't find anything, is not a sick person. He is a person who wants to be sick. Moreover, he wants to be sick in such a way that he can take control and self-medicate. He wants an imaginary illness, caused by imaginary toxins.

Now it's fair for you to stop me at this point and call me out on my claim that these toxic conditions are imaginary. I will now tell you why I say that, and then as always, you should judge for yourself.

Let's start with one of the more graphic detoxification methods, gruesomely pictured on web sites and in chain emails. It's a bowel cleansing pill, said to be herbal, which causes your intestines to produce long, rubbery, hideous looking snakes of bowel movements, which they call mucoid plaque. There are lots of pictures of these on the Internet, and sites that sell these pills are a great place to find them. Look at DrNatura.com, BlessedHerbs.com, and AriseAndShine.com, just for a start. Imagine how terrifying it would be to actually see one of those come out of your body. If you did, it would sure seem to confirm everything these web sites have warned about toxins building up in your intestines. But there's more to it. As it turns out, any professional con artist would be thoroughly impressed to learn the secrets of mucoid plaque (and, incidentally, the term mucoid plaque was invented by these sellers; there is no such actual medical condition). These pills consist mainly of bentonite, an absorbent, expanding clay similar to kitty litter. Combined with psyllium, used in the production of mucilage polymer, bentonite forms a rubbery cast of your intestines when taken internally, mixed of course with whatever else your body is excreting. Surprise, a giant rubbery snake of toxins in your toilet. Some brands use a different formula, consisting of guar gum and pectin, that accomplishes the same thing.

It's important to note that the only recorded instances of these "mucoid plaque" snakes in all of medical history come from the toilets of the victims of these cleansing pills. No gastroenterologist has ever encountered one in tens of millions of endoscopies, and no pathologist has ever found one during an autopsy. They do not exist until you take such a pill to form them. The pill creates the very condition that it claims to cure. And the results are so graphic and impressive that no victim would ever think to argue with the claim.

Victims, did I call them? Creating rubber casts of your bowels might be gross but I haven't seen that it's particularly dangerous, so why are they victims? A one month supply of these pills costs $88 from the web sites I mentioned. $88 for a few pennies worth of kitty litter in a pretty bottle promising herbal and organic cleansing. Yeah, they're victims.

It's already been widely reported that alternative practitioners who provide colon cleansing with tubes and liquids have killed a number of their customers by causing infections and perforated bowels, and for this reason the FDA has made it illegal to sell such equipment, except for use in medical colon cleansing to prepare for radiologic endoscopic examinations. There is no legally sold colon cleansing equipment approved for general well being or detoxification.

As usual, the alternative practitioners stay one step ahead of the law. There are a number of electrical foot bath products on the market. The idea is that you stick your feet in the bath of salt water, usually with some herbal or homeopathic additive, plug it in and switch it on, and soak your feet. After a while the water turns a sickly brown, and this is claimed to be the toxins that have been drawn out of your body through your feet. One tester found that his water turned brown even when he did not put his feet in. The reason is that electrodes in the water corrode via eletrolysis, putting enough oxidized iron into the water to turn it brown. When reporter Ben Goldacre published these results in the Guardian Unlimited online news, some of the marketers of these products actually changed their messaging to admit this was happening — but again, staying one step ahead — now claim that their product is not about detoxification, it's about balancing the body's energy fields: Another meaningless, untestable claim.

But detoxifying through the feet didn't end there. A relative newcomer to the detoxification market is Kinoki foot pads, available at BuyKinoki.com, and also now available from a number of knockoff companies. These are adhesive gauze patches that you stick to the sole of your foot at night, and they claim to "draw" "toxins" from your body. They also claim that all Japanese people have perfect health, and the reason is that they use Kinoki foot pads to detoxify their bodies, a secret they've been jealously guarding from medical science for hundreds of years. A foolish claim like this is demonstrably false on every level, and should raise a huge red flag to any critical reader.

Nowhere in any of their marketing materials do they say what these alleged toxins are, or what mechanism might cause them to move from your body into the adhesive pad. Kinoki foot pads contain unpublished amounts of vinegar, tourmaline, chitin, and other unspecified ingredients. Tourmaline is a semi-precious gemstone that's inert and not biologically reactive, so it has no plausible function. Chitin is a type of polymer used in gauze bandages and medical sutures, so naturally it's part of any gauze product. They probably mention it because some alternative practitioners believe that chitin is a "fat attractor", a pseudoscientific claim which has never been supported by any evidence or plausible hypothesis. I guess they hope that we will infer by extension that chitin also attracts "toxins" out of the body. Basically the Kinoki foot pads are gauze bandages with vinegar. Vinegar has many folk-wisdom uses when applied topically, such as treating acne, sunburn, warts, dandruff, and as a folk antibiotic. But one should use caution: Vinegar can cause chemical burns on infants, and the American Dietetic Association has tracked cases of home vinegar applications to the foot causing deep skin ulcers after only two hours.

The type of vinegar contained in these footpads is powdered wood vinegar. Powdered wood vinegar turns brownish black when exposed to moisture, such as the sweat produced by your skin when the adhesive pad is stuck on all night long. The color that appears on the pad has nothing to do with any "toxins" somehow being excreted from your skin.

Anyone interested in detoxifying their body might think about paying a little more attention to their body and less attention to the people trying to get their money. The body already has nature's most effective detoxification system. It's called the liver. The liver changes the chemical structure of foreign compounds so they can be filtered out of the blood by the kidneys, which then excrete them in the urine. I am left wondering why the alternative practitioners never mention this option to their customers. It's all-natural and proven effective. Is it ironic that the only people who will help you manage this all-natural option are the medical doctors? Certainly your naturopath won't. He wants to sell you some klunky half-legal hardware.

Why is it that so many people are more comfortable self-medicating for conditions that exist only in advertisements, than they are simply taking their doctor's advice? It's because doctors are burdened with the need to actually practice medicine. They won't hide bad news from you or make up easy answers to please you. But that's what people want: The easy answers promised by advertisements and alternative practitioners. They want the fantasy of being in complete personal control of what goes on inside their bodies. A doctor won't lie to you and say that a handful of herbal detoxification pills will cure anything that's wrong with you; but since that's the solution many people want, there's always someone willing to sell it.

31. MAGIC JEWELRY

Today we're going to try to heal our bodies through one of the simplest methods imaginable: We're simply going to slip on an ionized bracelet. It's going to correct our bodies' natural electrical fields, and anything that's wrong with us will be healed naturally. Does that sound a little suspicious to you? Well, it did to me too when I first heard it. But it sounds perfectly reasonable to a lot of people. Start paying attention and you'll notice that plenty of people still walk around wearing a Q-Ray bracelet. I have several friends who wear them. Once I asked one of them what it does, and he said he wasn't sure but he felt better whenever he wore it.

The Q-Ray bracelet is the best known and most popular of the so-called ionized bracelets on the market. It's actually a knockoff of the original, called the Bio-Ray bracelet, developed in 1973 by Spanish chiropractor Manuel Polo. Polo's idea was based on his belief that general wellness could be achieved through proper balancing of positive and negative ions in the body. Now, a vague claim like this might sound reasonable to an uninformed person, but run it past any electrical engineer. There's no plausible way that any type of bracelet could have such an effect, unless it's grounded. Walking across a carpet and scuffing your feet will have a dramatic effect on your body's static electrical charge (which is the technically correct way to refer to your body's "balance of positive and negative ions"), and this is unaffected by any jewelry you might be wearing. Being indoors or outdoors during a thunderstorm will produce the same difference. People riding in a car have a different static charge than people outside the car who are grounded. None of these conditions have ever been shown to have an affect on a person's general well-being, nor can they be mitigated by jewelry.

It's easy to poke holes in the technical claims made by the Q-Ray people, but their satisfied customers aren't interested. They feel better when they wear it, and that's all they need to know. Fortunately, that's a claim that someone finally decided to put to the test. The Mayo Clinic took 710 people and, in a fully blinded and randomized clinical trial, gave half of them Q-Ray ionized bracelets, and half of them identical looking "placebo" bracelets. As you can probably guess, there was no difference in the amount of pain reduction reported by either group.

I was curious to see exactly what the makers of Q-Ray claim on their web site. I went to QRay.com and found, quite anticlimactically, that they make no claims of any kind whatsoever. The bracelet's only characteristic that they discuss is its design. They don't even call it ionized any more. They simply sell them as metal bracelets, in various colors and styles ranging from $60 to $300. Turns out there's a very good reason for this. In 2003, the Federal Trade Commission nailed them for false advertising, and ordered them to stop. Unfortunately they didn't stop, and in 2006, the courts brought the hammer down hard. QT Inc., the makers of Q-Ray, were ordered to turn over $22.5 million in profits, and also provide $64.5 million in refunds. This ruling was upheld by the Seventh Circuit Court of Appeals in January 2008.

I wanted to see what these false claims had been, so I turned to the Internet researcher's favorite tool, the Wayback Machine at Archive.org, which archives old versions of web sites. But, I was thwarted once again. QT Inc. had had their site removed from the Wayback Machine's archives. Happily, there are even better archives out there. In this case, I found many of their claims preserved for the ages on Dr. Stephen Barrett's irreplaceable Quackwatch.org. Let's hear some of them. Cue the New Age music:

> *The Q-Ray bracelet is designed to achieve many of the same goals as traditional Chinese acupuncture. Acupuncture was developed to balance the body's Yin (negative ions) and Yang (positive ions), the two inseparable, complementary*

energies that permanently circulate in the human body. When these energies become unbalanced, the body's functioning is thought to be altered, which can be at least very annoying and at worst debilitating, depending on the size and nature of the energy imbalance. Oriental medicine, through acupuncture, is believed to regulate these two energies, discharging from the body excess positive ions and providing access to blocked negative ions, by stimulating meridian acupuncture points.

In the human body, which is electromagnetic by nature, biomagnetic alpha and beta waves circulate throughout the vital centers. When the flow is cut off and these alpha and beta waves become stagnant in one particular area of the body, bioelectrical alterations and ionic imbalances can result. Designed by Dr. Polo with polarized multi-metallic metals, the Q-Ray bracelet's circular form and spherical terminals offer low resistance to the bioelectrical conductibility of the alpha and beta waves, facilitating the discharge of excess positive ions or static electricity. Excess of positive ions is associated with poor nutrition, incorrect breathing, sedentary life style, and the use of electrical instruments or exposure to EMF (Electronic Magnetic Field). Loss of negative ions is associated with symptoms such as anxiety, stress, fear, hatred, and physical exhaustion.

Judge Frank Easterbrook aptly described these claims as "poppycock", "techno-babble", and "blather." He remarked that the "Defendants might as well have said: 'Beneficent creatures from the 17th Dimension use this bracelet as a beacon to locate people who need pain relief, and whisk them off to their home world every night to provide help in ways unknown to our science.'"

According to Dr. Barrett, one of QT Inc.'s owners testified in the 2006 trial that he could not define the term "ionization" but picked it because it was simple and easy to remember. The court concluded that his testimony on ionization was "contradictory and full of obfuscation". Did QT Inc. try to support their claims in court? No, they did not. Under oath, they testified that the only healthful effect produced by the Q-Ray was the placebo effect. That's right, sports fans: When under the gun, the Q-Ray's manufacturers stated that all the claims they've ever made about positive healthful effects from using their product are bogus. At best, it is a placebo. Just not a very useful one, according to the Mayo Clinic.

Dr. Barrett also archived claims from QT Inc. that the Q-Ray bracelets could restore health, relieve cancer pain, improve muscle flexibility, improve sports performance, restore energy, and provide other health benefits. A 2000 TV infomercial for Q-Ray stated: "When you have a severe injury or a chronic injury or a chronic problem like arthritis, you have build-up of positive ions. Wherever that is, you are going to have pain. In order to remove this pain, Q-Ray bracelet rips it right out of the body!" Needless to say, medical science has never found any correlation between pain or injuries, and build-ups of positive ions.

Q-Ray has not been completely neutered, though. As of this writing, a Google search returns a number of third party resellers of the bracelet still making all kinds of meaningless techno-babble claims. WellnessMarketer.com, for example, still calls it "ionized" and backhandedly implies that the Q-Ray has extremely powerful effects by warning that you should contact your doctor before wearing one if you're pregnant, that you should not allow the ends of the bracelet to contact each other, and that you should not wear a Q-Ray if you use a pacemaker or other medical device. Qbracelets.com says that the Q-Ray causes your body to "realign and balance its energies", and that it balances your body's yin and yang to "flood your system" with "increased amounts of natural pain relief compounds." ValueHealth2000.com says these are "the same bracelets you've seen in doctors' offices."

But there's little need for these marketers to rely on the Q-Ray. ValueHealth2000 also sells the Q-Link pendant, another knockoff that the FTC doesn't seem to have caught up with yet. They claim the Q-Link strengthens you against electromagnetic radiation, and also make the following statement that Judge Easterbrook would really love: "Doctors who tested the Q-Link Pendant with Sympathetic Resonance Technology™ found that it very quickly amplifies healthy energy states — and decreases energy drains caused by a wide variety of stressors." The FTC has, however, caught up with yet another knockoff company. They have a suit similar to the one against QT Inc. pending against the makers of an identical so-called "ionized" bracelet, the Balance Bracelet.

Unless you're trying to allege yet another pharmaceutical conspiracy, you have to think that any true healing powers of simple jewelry would have become a fundamental of medical science long ago, and would be first on your doctor's prescription pad. Just remember what we always recommend here: When you hear any claim that sounds far fetched or goes against your understanding of science or the natural world, you have very good reason to be skeptical.

32. World Trade Center 7: The Lies Come Crashing Down

Today we're going to point our skeptical eye, once again, at the events of September 11, specifically at World Trade Center 7, the building that collapsed after the twin towers for no apparent reason, in a manner consistent with a controlled demolition. We're entering the weird wild and wacky world of conspiracy theories, men in black, deceit, doubt, mistrust, and delusion. But on which side?

First let's be clear about what the two sides are, then we'll examine the evidence supporting each of them.

The conspiracy theory states that World Trade Center 7 was a controlled demolition, an intentional destruction of the building by our government. The evidence supporting this theory is threefold: First, the video of the collapse and the tidy distribution of the resultant debris appear consistent with known controlled demolitions. Second, photographs of the building before it collapsed showed little or no damage to cause a collapse. Third, fire alone cannot destroy a steel building, and so the cause must lie in high-energy explosives. A great deal more information is put forward by the supporters of this theory as evidence, but it's really only suppositions about proposed motives and observations of events perceived as unusual, and so is actually not testable evidence of a direct physical cause. This information includes government offices located in the building, the establishment of Giuliani's emergency management headquarters on the 23rd floor, and portions of the government's preliminary reports that openly stated that certain unknowns remained.

The competing theory is found in those very same government reports. The first, a preliminary report issued by the Federal Emergency Management Agency (FEMA) only eight months after the event, concluded that fires on the 5th through 7th floors caused the collapse, but infamously noted:

The specifics of the fires in WTC 7 and how they caused the building to collapse remain unknown at this time. Although the total diesel fuel on the premises contained massive potential energy, the best hypothesis has only a low probability of occurrence. Further research, investigation, and analyses are needed to resolve this issue.

Three years later, the National Institute of Standards and Technology (NIST) issued a working draft of the complete theory, scheduled to be finished in 2008. This report states that the building suffered two major failures, either of which could have been survived on its own, but not in combination. The first failure was severe damage to ten stories of the south side of the building, dramatically shown in a video from an ABC news helicopter, which destroyed several major columns. The second failure was the fire, fed in part by diesel generator fuel from high pressure tanks, which proceeded unfought for seven hours due to a lack of water pressure, and caused terminal weakening in the remaining columns that were already overloaded from the loss of the initial columns. Firefighters noted a growing bulge between the 10th and 13th floors and major structural creaking sounds, and finally evacuated. Two hours later, the east wall began to crack and bow. The east penthouse sank into the structure, and eight seconds later, the northeast corner fell, bringing the rest of the building down on top of it.

No evidence of any explosives were ever found, but the conspiracy theory states that this is because the government took away all the debris before it could be independently tested. Since it's normal for debris to be removed following any such destruction, this particular piece of information is too ambiguous to be given serious weight as proof of a conspiracy.

The claim that fire has never before destroyed a steel-framed building seems to hold up well, as it's hard to find a recent example of it. The reason is that modern building fires are always fought, they have sprinkler systems, and their steel is well insulated. Turn the clock back a few decades to World War II, when there was massive worldwide incendiary bombing of major cities, there were no sprinkler systems, and fire fighters had no hope of responding, there were many hundreds of steel framed buildings that were destroyed by fire. Not by bombs; by fire. The Edo Museum in Tokyo has preserved gnarled masses of giant girders twisted into knots by fire. London's Imperial War Museum has thousands of photographs of the same, and even a large collection of contemporary art depicting warped steel girders. Dresden's City Historical Museum also shows examples of steel girders from buildings that collapsed from fire, during that city's most infamous of all large-scale incendiary attacks. These museum collections all predate any alleged September 11 conspiracy.

There are three videos of the actual collapse that are of decent quality, and all show a collapse that appears reasonably consistent with what most laypeople have seen of controlled demolitions on television. The most obvious difference is that controlled demolitions start with multiple series of minor explosions distributed throughout the building to cut various support structures in a carefully planned sequence, followed a few seconds later by the charges to blow the key structural elements in a sequence designed to initiate the collapse in the desired direction. None of the videos of Building 7's collapse show any minor explosions. They simply show the top of the building begin to gracefully sag, as if it's made of clay, and then the whole thing drops. So while the manner of collapse may look superficially similar to a controlled demolition at first glance, a more careful examination shows critically important (and non-ambiguous) differences.

The neat, tidy arrangement of the debris of Building 7 is another characteristic of controlled demolitions that is claimed by the conspiracy theorists. WTC7.net states that "The pile was almost entirely within the footprint of the former building." In fact, Building 7's debris field was neither tidy nor well-contained within the footprint. The videos of the collapse are all from far away and show only the top portion of the building before it disappears behind the skyline. Lower down, the collapse become much more chaotic. Two nearby buildings were nearly destroyed by it. The Verizon Building suffered $1.4

billion in damage from the collapse of Building 7, but was able to be repaired. Manhattan Community College's Fiterman Hall building, however, was not so lucky, and suffered such major damage that it could not be saved. What remains of it is still being deconstructed piece by piece.

Could a building with such little apparent external damage collapse like this? The photos and videos on the conspiracy theory web sites are from other angles, and show only relatively minor, superficial damage to the building; and even the NIST has said the fire alone would probably would not have destroyed the building. But, let's not forget that Building 7 did have damage: Severe damage, a deep gouge cutting a quarter of the way through the building, ten floors high. Yet even if there was such extensive damage, argue the conspiracy theorists, that fact alone would invalidate the government report. Also from WTC7.net:

> *The alleged damage was asymmetric, confined to the tower's south side, and any weakening of the steelwork from fire exposure would also be asymmetric. Thus, even if the damage were sufficient to cause the whole building to collapse, it would have fallen over asymmetrically — toward the south.*

This claim forgets that nobody has said the damage alone was responsible for the collapse. According to the NIST report, the initial loss of the columns served only to transfer the building load to the remaining columns, thus exceeding their load bearing capacities, which then gave way after being adequately softened by the fire. In such a condition, the building would have insufficient support throughout. The east side, already sagging, dropped first and pulled the rest of the building down in a slightly diagonal collapse. The conspiracy theorists are correct in that the fall was not entirely symmetric, as it strayed enough to do the aforementioned damage to the Verizon and Manhattan Community College buildings. The conspiracy theorists have hardly proven that explosives are the only possible explanation for the collapse.

There's really nothing that's either mysterious or unexpected about the manner of Building 7's collapse. It was doomed by the damage, the diesel-fed fires, and the lack of firefighting capability. All the physical evidence, photographic evidence, and testimony of the firefighters is perfectly consistent with the government's official report. The conspiracy theory is supported by no evidence and is inconsistent with all of the events in the 7 hours preceding the collapse. The cause of Building 7's collapse is a question where very little critical analysis needs to be applied by a rational person. Judge the evidence for yourself.

33. MonaVie and Other "Superfruit" Juices

Have a seat and pour yourself a glass of the newest anti-aging megafad, superfruit juice. What is it? What does it claim to do for your body? How does it work? Is it really worth up to $50 per one-week supply?

One day I was logged into my Facebook account and noticed one of my friends had blogged on her page that she was sick, and she was "sure the reason was because she had run out of MonaVie". You can probably guess that this caught my skeptical eye. After all, these superfruit juices have only been available for a few years, and it's not like everybody was always sick throughout human history until they came on the market. Plus this was right around the new year, the height of the cold and flu season, and lots of people were sick. I wanted to know if it was really true that simply changing your morning beverage was the miracle cure to the common cold.

There are many of these superfruit juices for sale, and lots of them (like MonaVie) are sold through Amway-style multilevel marketing schemes. You've probably heard the question asked if you can make a better hamburger than McDonald's. Yes, of course you can. But: Can you build a better business than McDonald's? No. It's not about the hamburgers. McDonald's is not in the food business; they are in the real estate business. This same concept, at least at face value, appears to apply to MonaVie and its ilk. They are not in the fruit juice business; they are in the multilevel marketing business. Their product, like the Big Mac, is secondary to their business model. But let us not make a leap of logic and conclude that superfruit juices are the Chicken McNugget of fruit juice. Let's give them the benefit of the doubt, and listen to their specific claims.

Superfruit juices all build their claims about their product on the same central idea. They contain large amounts of antioxidants, which fight the free radicals that cause aging.

We're not going to blindly accept that without analysis. First, what really does cause aging? What the heck is a free radical? How is it affected by an antioxidant, and what the heck is an antioxidant? And, significantly: Do superfruit juices really contain beneficial levels of antioxidants? (With apologies to my Facebook friend, I didn't find any claims that these superfruit juices protect you from catching a cold — so save your $40-50/week if that's your goal.)

Free radicals are complicated. The 25¢ definition is a molecule with an unpaired electron that allows it to easily form a covalent bond with one of your good molecules, thus oxidizing it. This is one way that cells can be attacked, and this effect can and does lead to a number of age-related diseases.

At first glance, this makes the role of antioxidants obvious. Eliminate those oxidizing molecules, and help prevent age-related diseases. Right? Well, I'm sorry to say, not so fast. Human biochemistry is not as simple as the linguistic dichotomy of oxidation vs. antioxidant. It's extraordinarily complex. The oxidation from free radicals also has important benefits to the body: Converting fat into energy and attacking bacteria, just for a start.

Fighting disease consumes a huge portion of the scientific and medical budgets in the world, so a tremendous amount of research has been done into antioxidants. What these tests have found, overall, is that a certain amount of antioxidants is good, but too much is bad; but more significantly, the source of the antioxidants seems to have more importance than the amount. The primary phytochemicals that deliver antioxidants to the body are vitamin C, vitamin E, and beta-carotene. For the superfruit juices to fulfill their claims, they must therefore contain large amounts of these vitamins. The American Heart Association evaluated five studies of such superfruit juices for their efficacy in preventing cardiovascular disease, which is the main health claim about antioxidants. Of the five, two showed no effects, and three showed negative effects.

Dr. Stephen Barrett sums it up quite aptly in an article about antioxidants on his web site Quackwatch.org (which also lists this and many other clinical trials if you want to see for yourself):

> There is widespread scientific agreement that eating adequate amounts of fruits and vegetables can help lower the incidence of cardiovascular disease and certain cancers. With respect to antioxidants and other phytochemicals, the key question is whether supplementation has been proven to do more good than harm. So far, the answer is no, which is why the FDA will not permit any of these substances to be labeled or marketed with claims that they can prevent disease.

So now let's move on to our final question: Do these superfruit juices really contain significant amounts of antioxidants? They better, because they base their entire marketing campaign around this claim specifically.

Choice, the publication of the Australian Consumers Association, undertook a major study to answer this question in 2007. They bought virtually every superfruit juice that's commercially available. In their labs, they tested all of them for their total antioxidant capacity using the oxygen radical absorbance capacity assay test, laying out their methodology in detail — which you'll notice the promoters of these products never do.

As a baseline, *Choice* measured the total antioxidant capacity, or TAC, of a common apple — a Red Delicious Apple, to be precise — and got a reading of 5900. This number was then compared to the TAC measured from a daily serving of each superfruit juice.

The first type of superfruit juice tested was goji, a berry from Asia also known as the wolfberry. Servings of four different goji-based superfruit juices were found to have TAC measurements ranging from 570 to 2,025 for a product that is a 100% purée of the berry, in other words, from 10% to 34% the TAC of a common apple.

Next, they tested two brands of mangosteen superfruit juices. Mangosteen is claimed to have double the antioxidant capacity of goji. The results? Two two juices came in at 1,020 and 1,710, or 17% and 29% of the TAC of a common apple.

Next, they tested two brands of noni superfruit juices. Noni comes from Polynesia, and is frequently used in Hawaiian traditional medicine. The two brands measured 540 and 525, each about 9% the TAC of a common apple. In other words, a $7 cup of noni juice contains as much antioxidants as a thin 5¢ slice of apple.

Finally, *Choice* put açai to the test. The açai is a small purple berry from the Amazon, and Oprah calls it the "#1 food for anti-aging." Açai is the headliner ingredient in MonaVie, but since they do not disclose their formula, the percentage is unknown. *Choice* tested a similar product from RioLife containing 14% açai pulp. It measured almost as well as the best goji juice, with a TAC of 1,800, or about 31% as much as a common apple. (MonaVie, not surprisingly, did not allow their product to be included in the test.)

Choice also added several more common fruits to the mix. A single navel orange was found to have a TAC of 2,540. A cup of strawberries has 5,938. A cup of raspberries has a TAC of 6,058. And the overall winner was a cup of cultivated blueberries, with a total antioxidant capacity of 9,019.

But how can this be? *Choice* magazine found that the marketing literature says that goji berries have ten times, and açai berries six times, the antioxidant capacity of blueberries. Well, this might well be true. The difference is due to the fact that you're drinking a juice made from the fruit, you're not eating the whole fruit itself. For example, the mangosteen fruit has a huge amount of antioxidants and other nutrients. However, it's all contained within the inedible rind. The edible pulp of the fruit has only a negligible amount of either. This is how it's possible for the marketing claim to be, well, accurate if misleading; but the product itself to be devoid of the claimed benefits.

Superfruit juices may be good sources of antioxidants compared to, say, spaghetti or a cheeseburger; but if you want antioxidants, you'll get far more of them for about 1/100th the price by simply eating common fruit from the supermarket.

There's one final concern that critical minds should have with these superfruit juices, and with those who recommend them; and that's the conflict of interest inherent in a multi-level marketing scheme. Superfruit juices are available from many alternative practitioners (like chiropractors and naturopaths) who are not bound by any professional ethics, and even from some medical doctors who are. Each of them earns income on these sales through a multi-level marketing pyramid. When you read an article in a wellness newsletter touting the benefits of superfruits, the author makes money if you buy. When your friend recommends MonaVie to you, your friend makes money; and I bet your friend even tried to sell you on becoming a distributor — "because if you do, you can get it at a discount." Even your trusted personal trainer at the gym who makes vague anecdotal claims about superfruit's power is part of the pyramid. Be skeptical. Superfruit juices are a business model first; a salable product second; and a well-evidenced health product a distant third.

If you're truly curious about superfruit juices and want the truth, ask a source who has no financial interest in the product. Ask your medical doctor. You may find that he knows nothing about it; products like MonaVie that have no proven health value rarely find their way into medical or nutritional literature.

Now the default comment that I'm going to hear is the accusation that I'm on the payroll of Big Pharma, who are mortified that consumers will learn they can replace medical treatment and drugs with ponzi-pyramid juice drinks. Of course this makes no sense, since $40 a bottle is far more than it costs to buy apples for a week or simply eat a healthy diet like your doctor recommends; and if profit was their motivation, Big Pharma would be the first ones selling superfruit juices. But go ahead and make the comment anyway. Difference of opinion is what makes the world go around *(this comment not supported by scientific data).*

34. Water as an Alternative Fuel

Today we're going to pour a few drops of water into our car's fuel tank, and triple our mileage; we're going to electrolyze hydrogen from our municipal water supply and run our house; and with a cup of seawater — the most plentiful substance on earth — we're going to extract energy and solve the world's energy crisis. For today's topic is the use of simple water as an alternative fuel source.

If you've listened to the news at all over the past couple of years, you've probably heard several trumpeted headlines about energy being extracted from water. If you have an email account you've probably heard that the government and oil companies have been suppressing the fact that energy stored in water can power your automobile. Open any web browser and do a Google search, and you'll find claim after claim for energy from water. It's clean, it's free, it has no carbon emissions, and science just hasn't caught onto it yet because of some establishment conspiracy of silence.

The most recent one I heard of was a device to electrolyze water using the power from your car battery. The resulting gas is then inserted into your cylinder along with the fuel, dramatically increasing your engine's power, and thus reducing the need to burn gasoline. Since car engines create 12V electricity, there is an endless supply of juice to power the electrolysis. Fox News even broadcast a story about two guys using this same technology to power a Hummer for the US military, burning nothing but water. Sounds exciting, doesn't it?

Not too long ago on the news there was another claim. A retired engineer doing some sort of home-brewed cancer research found that seawater, when electrolyzed by radio waves, could be made to burn. The television reporters made all sorts of excited noise about this: Seawater is available everywhere for free, burning it produces almost no harmful emissions, and the heat from the reaction can be used to generate electricity or do just about anything else.

Can water really be used as a fuel? Has the solution to all our problems always been right under our very nose? Let me ask a different question: Is the idea that something so obvious could have gone unnoticed for so long absurd enough to warrant a healthy dose of skepticism?

Well, the short answer is yes, they do warrant skepticism; and no, they do not represent any new solution to any problem that nobody's ever thought of before. All of these miracle systems consume more energy than they create, and are reported by the television networks with no critical analysis of the bogus claims being made.

Let's start with the seawater guy. John Kanzius was tinkering with an idea he had to target cancer cells with metallic particles, and then blast them with radio frequencies to kill the cancer. During the course of his research, someone noticed condensation inside the test tube and they decided to try desalinating water. It worked; the intense radio waves caused water to electrolyze, releasing hydrogen. When ignited, this reaction could produce a continuous flame; and, of course, a flame can be used to do things like generate electricity. Different solutions and salinities produced different colors. Rustum Roy, a Penn State University chemist and a prominent advocate of holistic healing, called this electrolysis by radio waves "the most remarkable [discovery] in water science in 100 years." The electrical power required to generate the radio waves far exceeded the heat output of the flame, but that was never Kanzius' point. Somehow a warped description of this reached the media, who take no interest in a subject beyond a newsworthy angle. They irresponsibly reported that seawater was being made to burn and produce energy, completely neglecting all the important questions surrounding energy production. The media even took Roy's statement out of context and made it sound like he was proclaiming that this was the most remarkable *energy generation* discovery in 100 years, which is not what he said at all. In short, all Kanzius developed was an extraordinarily inefficient way to produce a small flame using tremendous amounts of electrical power from the grid. The water is not a fuel at all; it is merely a catalyst in one unusual method of converting radio waves to heat. Some people hear the explanation and say "Well, yeah, but it's brand new, they could work on it and make it more efficient, and then who knows the potential?" Would that that were so. Thermodynamics rules that even if the process could be made 100% efficient (which by itself is an absurd proposition), the heat output of the flame could never exceed the amount of energy coming from the electricity used to create the radio waves. By the way, John Kanzius' cancer research is still proceeding.

So how about the car engine thing? Use power from the battery — which is constantly being recharged by the engine — to electrolyze water, thus producing a volatile gas that can be added to the fuel mixture to substantially boost performance. The water tank needs to be refilled just like the gas tank does, and so in this case, the water is actually being used as fuel. Right? Not right. Welders who have heard about these devices generally fall onto the ground laughing when they hear it. These claims state that the water is converted into oxyhydrogen, the same gas used in water torches, and also known as Brown's gas. A water torch is a type of welding flame that uses oxyhydrogen as fuel. Oxyhydrogen is a gas that consists of hydrogen gas and oxygen gas in a 2:1 ratio, the same as water, but chemically separate from one another. Think of the space shuttle's main engines, which use liquid hydrogen mixed with an oxidizing agent. Recall the size of the explosion when the Challenger's main tank blew. Oxyhydrogen does have huge explosive potential, which is why it's such a great fuel for water torches. Water torches have been around for a long time, so there is nothing remotely new or inventive about this concept. It has never been of interest to automotive engineers because making the oxyhydrogen fuel consumes more energy than can be produced by burning it. Welding is not the art of energy efficiency, so this is not a problem for the welding industry. It would be a huge problem for the automotive industry, which cannot afford to spend more energy creating oxyhydrogen than could be produced burning it. The same goes for your car's engine. If your battery starts with a full charge, your car may indeed run more efficiently with one of these devices for a short time, until the battery is drained enough that the engine must take on the additional load of recharging it. And then there's that pesky law of thermodynamics again. It will never be possible to gain more energy burning the oxyhydrogen than it takes to create the oxyhydrogen. You can borrow energy from the burning gasoline to keep the reaction going, but now you are running less efficiently than you could under gasoline alone.

And so, alas for all such bogus claims of water as a fuel. Study them critically, and you'll find that they all represent net losses of energy. Be assured that engineers know more about physics than television reporters.

Now, it's important to note that some of these stories do have merit, if you're technically conversant enough to separate real science from bogus science. Bruce Crower, a lifelong tinkerer of racing engines in southern California, has invented what he calls the steam-o-lene engine. It's a conventional four-stroke internal combustion engine, with an additional two strokes tacked onto the end. Crower knows that the biggest waste byproduct of internal combustion engines is heat, and he decided to recover some of that by putting it to work in an additional power stroke. At the end of the engine's normal fourth stroke, which ejects the fuel exhaust, Crower injects a tiny amount of water. That water instantly flashes to steam inside the hot cylinder, creating a tremendously powerful fifth stroke. The sixth stroke ejects the steam, which goes to a passive condenser where it returns to water. Unlike the other examples we've discussed, Crower's system actually works. Crower understands that the water is not the fuel. The heat is the fuel. Water is simply a catalyst for converting that heat into kinetic energy. What's more, enough of the heat is recovered that you can eliminate the heavy radiator and cooling systems, and when running the engine is cool enough to touch with your bare hand.

So please, the next time you read about a new water fuel in the newspaper or hear about it on the news — which you probably won't have to wait long for — apply some skepticism. Find the data the reporters didn't want to dilute the impact of their headline. Demand a reasonable standard of evidence. Be skeptical.

35. SUPER-SIZED FAST FOOD PHOBIA

Join me for a cheeseburger and a Coke as we put our feet up, get grease all over ourselves, and examine the deeply-rooted pop culture belief that fast food is bad for you. And here's a thing of honey-mustard sauce to drink for dessert.

The questionable nutritional value of fast food, and of McDonald's in particular, came under its closest scrutiny when documentary filmmaker Morgan Spurlock released *Super Size Me* in 2004. The movie documented his own experience living exclusively on McDonald's food for 30 days. He averaged 5,000 calories a day, and when you consider that a Big Mac contains only 510 calories, you know that he was really packing it in. He super-sized every meal that was offered. Most dramatic was Spurlock's reported health problems. Not only did he gain 13% of his body weight, he also developed liver problems, depression and other psychological effects, and sexual dysfunction. *Super Size Me* also contained a large amount of editorial content about how McDonald's deliberately forces cheap, unhealthy food onto an unsuspecting public for profit.

Super Size Me was the most popular documentary of the year, and was nominated for an academy award. Its claims were generally accepted without critique by nearly everyone who watched it or even just heard about it. But this result was virtually guaranteed by Spurlock's choice of subject matter. McDonald's is probably the world's easiest target. It's always popular to be anticorporate; it's always popular to bash fast foods, and audiences are generally well predisposed to welcome any information that supports these concepts.

Spurlock's results were only presented in his movie. No actual data was published or subjected to any scrutiny or peer review. We have only his verbal statements to go on, plus the lines delivered onscreen by the doctor and nutritionist who performed in his movie. This is a Hollywood entertainment, it's not valid scientific data. However, for the sake of argument, my inclination is to give Spurlock the benefit of the doubt and accept his claims as valid, and accept the movie dialog as actual opinions of unbiased health professionals. From the perspective of responsible empiricism, that's a stretch, but I'm willing to do it. The problem is that Spurlock's results are highly deviant from other research on the same subject.

You see, Morgan Spurlock is not the only person to have ever tested fast-food-only diets, or even McDonald's-only diets. After his movie came out, many people repeated his experiment themselves, including a number of scientific institutions that applied controls and conducted the research in a scientific manner. At least three other documentary movies were made, *Bowling for Morgan, Portion Size Me,* and *Me and Mickey D*, in which the filmmakers lived exclusively on McDonald's food for 30 days but (unlike Spurlock) did not force themselves to overeat when they were not hungry. All filmmakers lost weight during the period and suffered no ill effects; and the subjects in *Portion Size Me*, which was scientifically controlled, also had improved cholesterol.

Most famously, Swedish scientist Fredrik Nyström conducted an experiment with seven students; only he upped the ante — considerably. Rather than Spurlock's 5,000 calories per day, Nyström's subjects were required to consume a measured 6,000 calories per day. The food was controlled to ensure that most of the calories were from saturated fats. The subjects were not allowed to exercise during the 30 days, also unlike Spurlock, who made sure that he walked a normal distance every day. Considering these differences, Nyström's subjects should have been considerably worse off than Spurlock was, but they weren't. They did all gain 5-15% extra body weight, and complained of feeling tired; but none suffered any other negative effects. There were no mysterious psychological problems, no strange conditions that "baffled the doctors". Nyström and his medical staff noted no dangerous changes at all. After his experiment, Nyström was asked his opinion of Spurlock's extreme reaction, especially his liver problems. Having never examined Spurlock, Nyström could only guess, but among two of his perfectly reasonable hypotheses were that

Spurlock may have had pre-existing undiagnosed liver problems; or that his normally vegetarian diet may have rendered his liver poorly prepared to suddenly deal with a diet high in carbohydrates and saturated fat, a problem that anyone eating a normal diet would not experience. Any cynic can also easily propose a third possibility, that Spurlock was simply trying to make as dramatic, engaging, and commercial a movie as he could, which is the goal of every filmmaker.

Public relations required McDonald's to respond to *Super Size Me*, and their response was fairly low key. They basically just agreed that it's best to eat a balanced diet, and stated that any actual ill effects experienced by Spurlock were more the result of force-feeding himself 5,000 calories a day for a month, than they were indicative of anything bad about McDonald's food. Way too much of any food is going to be bad for you.

That response suggests the next thing to look at. Is McDonald's food, and other fast food in general, actually bad for you? Dr. Dean Edell once took a call on his radio show from a woman whose teenage daughter ate a fast food hamburger every day. The woman was worried that her daughter would develop malnutrition. Quite the contrary, said Dr. Edell: She might gain weight if she ate a lot of them, but malnutrition is that last thing she should worry about. A hamburger is actually quite a balanced meal, rich with just about every nutrient. Add a slice of cheese and it even contains all four food groups. Fast food hamburgers are excellent sources of protein, calcium, and iron.

McDonald's hamburgers are not even as grossly calorific as most people probably think. Their biggest burger, the Double Quarter Pounder with Cheese, contains 740 calories. Three of those a day, which is more than anyone reasonably eats, still amounts to a good, healthy, slim 2,200 calorie diet for an adult. The real offenders on fast food menus are not the hamburgers at all, but the drinks and the fries, especially the milkshakes. Where Spurlock gained his weight was from the milkshakes. McDonald's 32-ounce Chocolate Triple Thick Shake packs 1,160 calories. Personally, I can't even imagine drinking a 32-ounce shake! A more common size, the 16-ounce, is 580 calories, or slightly more than a Big Mac. McDonald's biggest breakfast will also get you: The large Deluxe Breakfast delivers 1,140 calories. This may sound like a lot, but in fact it's not really much more than any average balanced breakfast.

By now you're saying "OK fine, McDonald's food may not be as high in calories as people think, but the real reason it's bad is that it's chock-full of trans-fats, sodium, saturated fats, and cholesterol." That would be bad indeed. The United States and Canada both use a system called the Dietary Reference Intake to establish ideal levels of nutrients. These four compounds listed have an ideal level of "as low as possible", except sodium. Ideally you should take 1500mg of sodium each day, and you should not take in more than 2300mg. McDonald's poster child of evil, the Big Mac, delivers 1040mg of sodium, about 2/3 of your daily ideal. Not a problem by itself, but don't eat three of them.

The Big Mac delivers 10g of saturated fat, which is 10g more than you want; but realistically it's virtually impossible to get zero. The Center for Disease Control and the World Health Organization recommend that you keep your saturated fat intake under 7% of your daily caloric intake, and the Big Mac fulfills half of that. So, in short, two Big Macs a day maxes out your recommended safe levels of saturated fat.

The Big Mac's 75mg of cholesterol represents 1/4 of the CDC and World Health Organization's daily recommended maximum. I'm not going to eat four of them a day, so that's not a problem.

Finally, the scariest mugshot on the CDC's Ten Most Wanted poster: trans-fats. Beginning in 2003 with some high-profile class action lawsuits filed against major food producers, the fast food restaurant chains have all pledged to switch to cooking oils free of trans-fats. Some have completed this, others, including McDonald's, are still completing the switch. But although it's possible to eliminate the addition of trans-fats to fried foods, some foods, like meat and some vegetables, contain naturally occurring trans-fat. 2-5% of the fat in livestock is trans-fat. Whether you order a Big Mac or barbecue your own organic filet mignon, you're getting trans-fat. McDonald's doesn't add it, and your neighborhood butcher has no way of reducing it. A big Mac (or any comparable meat of the same quantity) contains 1.5g of trans-fat, which is more than you want, but only about 8% of the daily amount the World Health Organization says you really, really need to keep it under. Eight percent — the Big Mac is hardly the monster it's made out to be.

So eat up, and I'll see you at the drive-thru.

36. Despicable Vulture Scumbags

Today I'm going to depart from the usual format and answer a single email that I received. You get a lot of email when you do a show like the Skeptoid podcast — most of it praise or thanks for helping a friend; a bit of it negative, which is usually both obscene and anonymous; and once in a while something that really gets my attention. The following email was of this latter variety:

> *I am writing you because I am upset, confused and angry... very, very angry.*
>
> *My good friend was just diagnosed with amyotrophic lateral sclerosis (ALS) and it is heartbreaking. He has three very young children. I had lunch with my friend today and listened to him tearfully speak of his fear for his wife and children's future, his rapidly deteriorating motor skills and the agony he will shortly endure. He is scared, as anybody would be.*
>
> *His doctors say that there is little hope for him.*
>
> *Unfortunately, he has been given the advice, perhaps by someone who wished him well, to undertake homeopathic therapies, specifically some warped device called the Baar Wet Cell Battery. To say that I am angered by this obvious attempt to cash in on my friend's fear and horrible condition is a monumental understatement. The device costs nearly $200 a month and doesn't even include instructions, which are an additional $15.*
>
> *It's obvious that the $200 would be better spent on his family or even on research to discover a treatment or cure. This device is a crime against humanity, perhaps not physically, but philosophically. The people who perpetrate it should be incarcerated.*
>
> *I have made the very difficult decision to not try to dissuade my friend from using this device as for now it is the*

only hope he is clinging to. Perhaps you have some advice you would be willing to give me and others in similar situations? Perhaps you can change my mind and suggest what to do...

When I got this email, it really affected me. I could easily imagine myself in the friend's position, and in the listener's position as well. Sadly neither position is unique. There are many such small personal tragedies around the world, and many people asking your same questions.

Amyotrophic lateral sclerosis, or ALS, is better known as Lou Gehrig's disease, and maybe best known these days as Stephen Hawking's disease. The sclerosis is a degenerative hardening of nerve cells in the brain and spinal column, resulting in painless weakness and atrophy of muscles, often leading to complete paralysis. Only one cause is known, a mutation responsible for only a few percent of cases. The average life expectancy is only two to five years after diagnosis, though five percent live 20 years. Stephen Hawking has survived 45 years so far. Over 5,000 people develop ALS in the United States each year, most of them white males over 40. In some cases, the progression stops, and a few cases have even reversed. ALS progresses differently in virtually every patient, and is thus exceedingly difficult to diagnose. There is only one approved treatment, the drug Rilutek, which has been shown to slow the progression of ALS in some cases and prolong life by a few months. No other treatments so far have been found to have any beneficial effect, though a number of clinical trials for other drugs are ongoing.

The use of a wet cell battery in New Age healing is the invention of Edgar Cayce, an early American celebrity psychic from the early 20th century. Cayce was best known for giving psychic readings on the sick, and developed a following of believers in psychic healing which still persists to this day. Virginia's Association for Research and Enlightenment, which claims tens of thousands of members, still promotes spiritual and psychic healing through methods developed by Cayce. One of these is the use of a wet cell battery. To use it, you mix chemicals in a bucket to create a weak battery. Next you mix the special "medicines", gold, silver, spirits, or iodine, in a secondary jar. A wire loops through these "medicines" and you apply the electrodes to your spine. You're supposed to do this for 30 minutes a day, and Cayce said months to years of daily application is required to get results. The claimed mechanism is "to introduce energy and medicinal vibrations directly into the body". There is no hypothesis behind the device suggesting how or why it might be useful.

The battery kit itself is $200 from this company Baar, but then there's a gigantic list of accessories and chemicals that you'll need a steady stream of, and these appear to account for the ongoing costs mentioned in the email.

The term "medicinal vibrations" is linguistic nonsense with no scientific or medical counterpart, so it was difficult to search for any research on this. I was not able to find any clinical trials or any research of any kind on a reputable source such as PubMed, into the efficacy of wet cell batteries for either wellness or treating disease. The Cayce web page for the battery makes no claims about its powers, although there is a long list of personal testimonials from customers of the Baar wet cell battery for all kinds of diverse conditions like fatigue, inner harmony, repolarization, and cold hands. Oddly, the authors of these testimonials managed to realize their positive results in far less than the "months to years" that Cayce said was required, often just days. Baar does point out that the device is not intended to diagnose, treat, cure or prevent any disease, and makes no claims about its usefulness. Apparently they leave that to people like whoever it was who recommended this product to your friend. Nevertheless, if Baar is knowingly selling this device to someone who believes it may treat ALS, they are indeed despicable vulture scumbags.

I'd like to reprint the response of Dr. Stephen Barrett, who runs Quackwatch, to one of his frequently asked questions: "Isn't it better to at least try an unproven alternative therapy, than to simply give up, roll over and die?" Dr. Barrett says:

> *I recommend taking whatever steps are needed to determine the accuracy of the 'terminally ill' prognosis. If it is*

correct, I would recommend spending the remaining time in the most productive way. In my own case, I would place my affairs in order, and continue to write about the topics I believe are most important. I would not waste 10 cents or 10 minutes looking for something that does not exist.

There's one thing that I would add to Dr. Barrett's suggestion, and it's a matter of personal choice. I'd probably participate in clinical trials, though I'd do it with the understanding that they are rarely successful, and with an awareness of potential side effects. Future ALS sufferers may benefit from my participation. Clinical trials are always free to the participants, and that's another important distinction between proper medical trials and quack treatments. The purveyors of quack nonsense like the wet cell battery always, always charge money. The ALS Association web site maintains a large list of clinical trials in which your friend might choose to participate. At best, he may find relief if one of these treatments pans out. At worst, his participation will further our knowledge of ALS and get us a step closer to the day when it's treatable.

ALS is no picnic, but it's also not the end of the world. Your friend has three small children who are always going to look up to him as their guiding light. He has their youth sports leagues to watch, their high school graduations, their first dates, their first broken hearts, their weddings. He has friends like you watching out for him, hanging out with him, taking him to the movies. How much of that will he be around to enjoy? Nobody knows. But then, you or I might get hit by a bus tomorrow. In that sense we all have the same concerns, and we can't all stop living our lives just because we don't know what tomorrow brings.

It is the nature of human beings to come and go. Passing on is neither unnatural nor unexpected. It's the way the world works. Death is not the culprit. The obscenity here is that someone is getting $400 a month richer off of your friend's tragedy. Someone who never went to medical school. Someone who probably couldn't tell you anything about the mechanisms of ALS. Someone who, I promise you, has never successfully treated ALS, and who has no plausible chance of ever doing so. You've said you're reluctant to dash your friend's hopes. But hope does not come in the form of a worthless piece of pseudoscience. Hope comes from a realistic chance, no matter how small — get him into a clinical trial if he wants to work the problem. The demonstrably false hope sold by Baar and other Edgar Cayce followers is a ripoff and a lie. Your friend, and everyone like him who needs a chance and a break and some hope, deserves better.

You asked my advice, so I'll give it, controversial though it may be. Having ALS doesn't give your friend any special dispensation to cause additional grief to his family and friends. He still has obligations to them, he's not above being wrong, and he's not the only person affected by this. Tell him he's making a bad situation worse for those close to him. Make sure he has the tools to make an informed decision. Share your thoughts with his other family and friends. Be part of the solution and don't enable the problem.

37. Can You Hear The Hum?

Close the windows, turn off the electricity, and be very quiet: We're listening for the Hum, a worldwide phenomenon in which a distant rumbling sound can be heard in some places by some people. No single cause has ever been found. The Hum is infamous in some of its most noted locations: The Taos Hum in New Mexico, The Bristol Hum in England, the Auckland Hum in New Zealand, the Kokomo Hum in Indiana. In these places, some 2-10% of the population can hear the rumble. It's described as sounding like a distant diesel engine idling. Some people hear it better outdoors; some people hear it better indoors; some people hear it higher up on the second story and others lower down in the basement. In some places, more men hear it than women. In others, more women hear it. Some Hums are heard more often by older people, and some by younger people. For some people, earplugs help — indicating that it's an actual audible sound; for others, they don't — indicating that it's not. Explanations ranging from insect noise to meteors to secret government projects abound, but no explanation is satisfying.

So what exactly does this Hum sound like? A number of people have made synthesized versions of the Hum with the cooperation of sufferers, sort of like an audible police sketch of a suspect. Dr. Tom Moir in New Zealand has done some research on the Auckland Hum, and has collected an actual audio recording. It's really low frequency, and the best way to describe it would be something like a big diesel engine idling far away. Basically, a rumble.

Some people I spoke with did cast doubt on the authenticity of Moir's recording, saying that nobody has ever successfully managed to record the Hum, and that his recording sounds identical to some of the synthesized versions out there. However, when presented for purely illustrative purposes, this recording does give an accurate representation of the general consensus for what the Hum sounds like. In response to my email inquiry, Dr. Moir replied:

> *The recording on my web page is for real.* Having said that, this does not imply some great mystery since very low frequency sound can travel for vast distances.

If the Hum can be recorded by audio equipment, that proves that it's an actual audio phenomenon. But others have failed to record anything, and have put forth other possible explanations. Dr. David Deming of the University of Oklahoma has probably done the most scholarly research of the Hum, though he's quite forthright in the lack of testable evidence. Hum research has had, thus far, to rely heavily on anecdotal reports and personal stories. But Dr. Deming has managed to conclude that the most probable explanation is that some people have been found to be able to hear radio waves.

Now before you spring for your tinfoil hat, allow me to present a snippet from the conclusion of the best paper on this phenomenon, Human Auditory Perception of Pulsed Radiofrequency Energy, by Drs. Joe Elder and C.K. Chou of the Motorola Florida Research Laboratories:

> *Human perception of pulses of RF radiation is a well-established phenomenon that is not an adverse effect. RF-induced sounds are similar to other common sounds such as a click, buzz, hiss, knock or chirp. Furthermore, the phenomenon can be characterized as the perception of subtle sounds because, in general, a quiet environment is required for the sounds to be heard. To hear the sounds, individuals must be capable of hearing high frequency acoustic waves in the kHz range and the exposure to pulsed RF fields must be in the MHz range. The experimental weight-of-evidence does not support direct stimulation of the central nervous system by RF pulses.*

I did not find this research to be a convincing explanation for the Hum, and the reason is that the perceived sound that subjects reported was radically different from descriptions of the Hum. Apparently, in these cases where powerful RF pulses can induce a perceived sound in some humans, the frequency of the perceived sound is related to the size of the head and mass of the brain of the listener; it is not related to whatever signal may be contained in the RF. Adult humans who can perceive RF will seem to hear a sound around 13 kHz. That's a really high pitched sound; too high for a lot of people to hear. No matter how you break up a 13 kHz tone into clicks, pops, or chirps, it's never going to sound anything like descriptions of the Hum. Thus, the evidence we have about humans hearing sounds caused by RF is that it's a very poor candidate for the Hum.

And just what might these radio sources be? The most frequently blamed suspect is the US government's High Frequency Active Auroral Research Program (HAARP) in Alaska. This is a research project that transmits RF straight up into the ionosphere, at approximately 1/10,000 the power of the sun's normal electromagnetic radiation. So far it has been able to produce a tiny artificial aurora, detectable by sensitive instruments but not by the naked eye; and also Very Low Frequency (VLF) waves at .1 Hz, which are otherwise difficult to create. Mentioning HAARP and the Hum in the same sentence appears to imply some kind of connection, and of course any government technology project raises suspicion among the paranoid; but I see no plausible connection between the two. There's been no correlation between HAARP and the Hum in either time or space. Reports from Hum sufferers did not increase when HAARP began only recently, and localized Hum phenomena have never been near the HAARP site either before or since it began. And, as discussed previously, the potential acoustic effects of RF radiation are completely dissimilar from the Hum.

Others blame cell phone networks or LORAN, the radio-based predecessor to the Global Positioning System. These candidates have the same evidential problems as HAARP and their only real support comes from the crowd that promotes the pseudoscience of modern electromagnetic fields as health hazards.

Mass hysteria has also been put forward as a possible cause. If the Hum is some kind of hysteria, it's certainly not a mass one. Very few people hear the Hum, even in the hotbed areas. Psychoacoustics and auditory hallucinations are not unheard of, and have been correlated with other physiological effects of stress. I did a fair amount of searching around the web to see if I could find any cases of Hum sufferers being treated with psychotherapy or other stress reduction, but did not find anything; so there does not yet appear to be any data supporting this hypothesis. But, given the total number of people who have experienced the Hum over the years, it seems probable that at least some of those cases could be explained by psychophysiology.

If you go to your doctor to complain about the Hum, the most likely diagnosis you'll get is tinnitus. This is the ringing in the ears that everyone gets at some point, and is often associated with ear infections, tube blockages or even head injuries. I've had this probably about as much as most people, and to me it sounds nothing like the Hum. However, by yawning or by tightening the tensor tympani muscle inside my ear, I can induce a loud, low-frequency rumble. It's hard to describe exactly how I do it, but I can make it last for maybe 30 or 40 seconds before the muscle fatigues. When I do this, it sounds exactly like descriptions of the Hum. It's also gotten stuck a few times when I've had a cold or blown my nose too hard, and when it goes by itself, it tickles and is really annoying, and I end up with this rumble in my head for a while. It's not hard to think that some people may have this condition chronically, and since this is the exact sound described by Hum sufferers, it's virtually certain that some variation on this condition is the explanation for some of them.

The city of Kokomo, Indiana hired a firm, Acentech Incorporated, to find the source of the Kokomo Hum and suggest solutions. The lead investigator, James P. Cowan, did find two sources of industrial noise that were likely candidates: Some cooling fans at the local DaimlerChrysler factory emitting a 36 Hz tone, and an air compressor at the Hayes International plant emitting a 10 Hz tone. These were alleviated, but complaints did not cease altogether. Cowan's investigation was thorough and he did conclude that there was probably something else causing at least some of these complaints.

So how do you wrap up a question like the Hum? When you assemble all the research and reports, you get a lot of

footnotes, some data, some hypotheses, but mostly a giant pile of question marks. I think it does all lead to one conclusion that is pretty certain: There is no Hum. At least, not a single worldwide phenomenon that we can lump together and call the Hum. There are many people all over the world who perceive a low rumble under certain conditions. Many of them are probably hearing an actual audible sound from some relatively mundane, yet undiscovered, source. Some are probably suffering from a problem with tinnitus or the tensor tympani muscle.

Some are probably experiencing an auditory hallucination. Some may be hearing an undiscovered geophysical phenomenon. And there are probably some hearing something from a cause that nobody has even hypothesized about yet. But there are also many people experiencing similar things: Different types of sounds, strange lights, unexplained feelings. We don't call all of those the Hum too. Whatever the various causes of these peoples' experiences is, it seems clear that there is no one quantifiable Hum that adequately explains all these diverse reports. Thus, anyone doing "Hum research" is really pursuing something that probably does not exist. Yes, it's possible that most of these cases share the same cause, but it's much more likely that very few of them do.

38. THE "TERROR" OF NUCLEAR POWER

Let's have a seat at Homer Simpson's control panel, chow down on some donuts, and nap away into oblivion while blinking lights and buzzers warn of impending doom and that glowing green bar of uranium that fell into our trousers. Today we're going to examine the popular notions about nuclear power. Specifically, if xenophobia had not killed nuclear power in the United States in the late 1970's, there's a good chance that we might have all been driving electric cars for the past 20 years; and uncounted billions of tons of carbon dioxide would never been sucked out of the ground, burned in power plants, and exhausted into our atmosphere.

So let's state the obvious. The immediate reaction to that statement is "OK, that may be true, but look at all the new problems we'd have created with Chernobyl-type disasters and lethal nuclear waste." Fair enough, and important questions, to be sure. Let's start with a quick primer on the various types of nuclear reactors.

So-called Generation I reactors were the early prototypes developed by many nations, and actually placed into production in a few cases. Generation I reactors were characterized by fundamentally unsafe designs, and kludged layers of afterthought safety systems. When most nuclear nations began deploying commercial reactors, they were usually of Generation II design. Generation II reactors were significantly improved, but these changes were primarily evolutionary. Most of the commercial plants in operation in the United States are Generation II designs. A little over ten years ago, Generation III designs began appearing in some of the world's most advanced nuclear nations. Generation III reactors incorporate not only evolutionary improvements, but also revolutionary changes such as fuel cycles that result in much less nuclear waste; reduced capacity for the creation of weapons-grade plutonium; and passive safety designs wherein the reaction cannot be sustained in the event of a problem and the system effectively shuts itself down, by virtue of its basic design. The newest plants being designed for commercial use are called Generation III+, which incorporate all the newest knowledge from operating Generation III designs. If a new reactor was approved and built in the United States today, it would be a Generation III+ design. Even if every plant employee keeled over with a heart attack, neither a Chernobyl nor a Three Mile Island type accident would be possible (within reason); the systems are fundamentally redesigned so that the reaction cannot be sustained if things go outside the parameters.

The Idaho National Laboratory is the United States' primary advanced reactor research facility, and they've outlined six new reactor types to be developed for Generation IV. The designs take everything to a new level: Lower cost, safer designs, near-total elimination of nuclear waste, and reduced risk of nuclear weapons proliferation. There are also Generation V reactors in the ether, but these are primarily the domain of late-night rumination sessions at the lab, fueled by tequila and pot.

Then there's fusion power, which is everyone's ultimate goal. Fusion reactors have the profound advantages of using simple tritium or deuterium for fuel, producing no significant waste, and absolute safety since if anything goes even slightly off-kilter, the plasma disappears and you have no reaction. It's the ultimate in cheap, clean, safe, renewable energy, despite gross misunderstandings of the technology expressed by Greenpeace and other factions. The first operational tokamak fusion reactor for research is being built by the international ITER consortium in France and is expected to come online in 2016.

So you can probably guess that Three Mile Island was probably not the newest and safest design, and you'd be right. It was a Generation II design. It was the first and only significant commercial nuclear accident in American history (there have also been a number of minor, harmless accidents in military reactors). A broken valve caused coolant to leak into a containment facility designed for that purpose, raising the temperature of the core and causing a partial meltdown. Despite significant confusion on the part of the operators (this being their first experience with an accident), and a somewhat lengthy chain of errors and misunderstandings, everything eventually worked out just as it should. There were no deaths or injuries, and despite 25,000 people living within five miles of the plant, nobody was exposed to any radiation worse than a single chest x-ray. All the studies predict zero cases of future cancer, despite ongoing lawsuits that the courts continue to find to be without merit. With proper perspective, Three Mile Island can (and should) be characterized as a shining example of how well the safety systems work, even in the face of human error and old-fashioned reactor design.

But that's not the way it was perceived. By an unfortunate coincidence, Jane Fonda's movie *The China Syndrome* about a nuclear accident came out only twelve days before Three Mile Island. The Cold War with Brezhnev was in full force and the words "nuclear accident" were simply too much for a scientifically uninformed public. Three Mile Island became the first nail in the coffin of American nuclear power.

Seven years later in 1986, things got much worse. Chernobyl was suffering from inadequate funding. Much basic maintenance had never been performed. It had only a skeleton crew, nearly all of whom were untrained workers from the local coal mine. The only manager with nuclear plant experience had been a worker installing small reactors on board Soviet submarines. Some genius decided to run a risky test of a type that no experienced nuclear engineer would ever gamble on. The test was to shut down the water pumps, which must run constantly in that type of reactor; and then find out whether the turbines, spinning on their momentum alone, had enough energy to restart and run the pumps during the forty-second delay before the backup diesel generators would kick in. The test was so risky that one faction within the plant deliberately disconnected some backup systems, trying to make the test too dangerous to attempt. The test was run anyway. It didn't work, the pumps couldn't keep up, the graphite core caught fire, the coal miners couldn't find any shovels so they didn't know what to do, and the reactor exploded. If you think I'm exaggerating this, there are extensive resources both online and in print, if you really want the hairy truth. In this short space I'm probably not even giving you ten percent of what a travesty this was — I'm tempted to call it a joke but it's so not funny. For example, they scheduled this right in the middle of a shift change, and the new workers coming in didn't even know what was going on.

Two people died that day, and some 30 to 60 people were dead within three months. Predictions of eventual cancer deaths caused by the radiation run from 1,000 to 4,000. And, of course, the damage to the local environment is extensive and difficult to estimate. The terror of a radiation cloud blowing across Europe was the second nail in the coffin of American nuclear power.

Not only was Chernobyl a monumental failure of the human element, the plant was a Generation I design, specifically an RBMK reactor, which is generally regarded as the least safe reactor type ever built. One design flaw is that the core used combustible graphite, and this distinction is the main reason that Chernobyl-type disasters are not possible in most reactors around the world. Only a very few Generation I designs are still in use, all in the former Soviet Union, and all have been retrofitted with improvements intended to prevent this type of accident. Other nations have long been lobbying for the closure of these reactors, and rightfully so.

How do the dangers of nuclear energy compare to the dangers of fossil fuel energy? A report in the Journal of the American Medical Association found that some 50,000-100,000 Americans die each year from lung cancer caused by particulate air pollution, the biggest cause of which is coal-burning power plants in the midwest and east. Even taking the maximum predicted death toll from Chernobyl, we would need a Chernobyl-sized accident every three weeks to make nuclear power as deadly as coal and oil already is. Shall I repeat that? If the world was filled with Generation I reactors run by feuding coal miners, we would need a worst-case scenario every three weeks just to match the US death toll we've imposed upon ourselves by clinging to our current fossil fuel system. Next time you see activists cheering the defeat of nuclear power in the US, realize that a healthy environment and saving lives are clearly not their priorities.

Well, maybe to them it's more about the future of the planet than about saving lives today. Maybe they just don't want to see high-level nuclear waste created that's going to poison the planet for tens of thousands of years. I can see that. But here's the problem with that logic: The plants we're designing now produce less waste than ever. Some on the drawing board produce none at all. We've already created most of the waste that we ever will. It already exists. It's out there. Lobbying against future cleaner plants won't make the existing waste go away. It's out there now in temporary facilities in neighborhoods all across the country, way more vulnerable than it would be in proper permanent storage in Yucca Mountain.

Opponents say that Yucca Mountain is geologically unstable or otherwise too hazardous, so the waste might leak out. Well, trust me: The location of the Yucca Mountain site was one of the most lengthy and expensive decisions the government ever made. What do you think they were doing with all that time and money, picking their noses? Well, it was a government program, so a large part of the time and budget probably was spent on nose mining. Nevertheless, this was one of the most scrutinized decisions ever made. Environmentally speaking it's as good a site as we could hope for. If you're concerned about it, go to a neutral and reliable source and research it personally. From every scrap of reason I can muster, environmentalists should be Yucca Mountain's #1 fans. I can't imagine why they prefer to leave the waste out where it is now, unless they are driven more by ideology than by science. Who would have thought that?

There is a safe and clean solution to our energy crisis, gasoline prices, and global warming. It's the latest generation nuclear reactor.

39. Apocalypse 2012

Abandon all your possessions and run for the hills: It has been foretold that the world is coming to an end sooner than you think, in the year 2012. It seems that you can't pick up any newspaper or magazine without reading that the apocalypse is almost upon us.

What really is going to happen in 2012? Asteroid 433 Eros is going to pass within 17 million miles of the Earth in January; the United States will hand over control of the Korean military back to the Koreans in April; there will be an annular solar eclipse in May and a solar transit of Venus in June; the Summer Olympics will take place in London; the Earth's population will officially pass 7 billion people in October; the United States will elect a new President in November; construction of the new Freedom Tower will be complete in New York City; the sun will flip its magnetic poles as it does at the end of every 11-year sunspot cycle; and, as I'm sure you've heard by now, the Mayan calendar completes its 5,125 year cycle, presumably portending the End of Days.

Mayans had three calendars. They had a solar calendar that was 365 days long, and a ceremonial calendar that was 260 days long. These two calendars would synchronize every 52 years. To measure longer time periods, they developed the "long count" calendar, which expressed dates as a series of five numbers, each less than twenty; something like the way we measure minutes and seconds as a series of two numbers each less than sixty. And, just in case this might seem too simple, for some reason the second to last number was always less than eighteen. The first day in the Mayan long count calendar was expressed as 0.0.0.0.0, and by our calendar, this was August 11, 3114 BC. Every 144,000 days (or about every 395 years, which they called a baktun), the first number would increment, and a new baktun would start. Recall how we all got to enjoy the excitement on the millennium of watching the digital displays roll over from

12/31/1999 to 1/1/2000? Well, that's what's going to happen on December 21, 2012 to the Mayan calendar. It's going to roll over from 12.19.19.17.19 to 13.0.0.0.0, just as it has done each of the previous twelve baktuns. There's no archaeological or historical evidence that the Mayans themselves expected anything other than a New Year's Eve party to happen on this date: Claims that this rollover represents a Mayan prediction of

the end of the world appear to be a modern pop-culture invention. It's true that the Mayan carvings of their calendar only depicted 13 baktuns, but what did you expect them to do? Carve an infinitely long calendar every time they wanted to express a date? The explanation could be as simple as they didn't expect people in the 21st century to still be obsessed with their archaic calendar.

Another story predicting doom in 2012 says that a new planet, variously described as Planet X, a planet/comet (which makes no sense), or the planet "Nibiru" is going to pass so close to the Earth as to cause earthquakes and tidal waves and all kinds of destruction, possibly even flipping the Earth completely upside down. This is an urban legend that's been around for a long time, but for most of the story's history, this was supposed to happen in May of 2003, as any Internet search for "Planet X" will reveal. Apparently what happened is that the Planet X advocates, perhaps embarrassed or disappointed that 2003 passed without incident, heard about the much more popular Mayan calendar story, and decided that 2012 is close enough to 2003 that it must be the correct date and that the Planet X destruction is probably what the Mayans were foretelling. The Planet X legend got started by misinterpretations of astronomical observations combined with an ancient Sumerian carving that has been erroneously interpreted to depict a solar system with ten planets. Why the craftsmen who made carvings in ancient Sumeria should be presumed to have planetary knowledge superior to that of modern astronomy is not convincingly argued. Of course, there is no Planet X, comet, planetoid, or otherwise. If you're interested in all of the actual science behind the Planet X story, there's no better source than Phil Plait's *Bad Astronomy* blog, which goes into all the facts, rumors, and sources in detail.

Here's one more reason people are frightened about 2012. About 500 years ago, Copernicus confirmed what Hipparchus had observed in 2200 BC: that the axis of the Earth, which leans over at 23.5°, completes one full rotation every 25,765 years. This means that in 12,000 years, Christmas will come to Australia in winter and the northern hemisphere will depict Santa in Bermuda shorts. Astrologers call this period a Great Year, and they divide it into 12 Great Months or astrological "ages", each about 2,147 years long. Each age corresponds to one of the signs of the zodiac. We are currently in the Age of Pisces, and like the song says, we're soon going to enter the Age of Aquarius. According to modern official delineations of the edges of the constellations, we'll move into the new age in the year 2600. But there's some disagreement, and some astrologers place it at 2595, 2654, or 2638. A few put it much earlier, as soon as 2150 or even 2062. However, once the news of the

Mayan calendar broke, a large segment of the astrological community abandoned the official constellation definitions and stated that the Age of Aquarius will begin in 2012. So, you can call this a third major reason why the world will end in 2012, but you have to be awful loose with your astrology, and you also have to think of some reason why the dawning of the Age of Aquarius might bring on the end of the world. I have not found any plausible claims for how it might have this effect.

So that's a lot of reasons, weak though they might be, to predict that the we're all going to die in 2012. However, there's one significant fact that the 2012 doomsayers all seem to forget: Despite all the various 2012-ish predictions for the end of the world, there are far more stories of apocalypse with different dates. For example, popular interpretations of Nostradamus found predictions for the end of the world in July of 1999, December of 1999, June of 2002, and October of 2005. It's also been said that his writings could mean the dead will rise from their graves in either 2000, 2007, or the year 7000. Nostradamus never said anything about 2012.

Many Protestant Christians believe that the end of the world will come in the form of what they call the Rapture, when the righteous will all be whisked away to heaven. Shakers believed the Rapture would come in 1792. Seventh Day Adventists first calculated it would happen in 1843, then when nothing happened, they found an error in their calculations and corrected it to 1844. The Jehovah's Witnesses made firm predictions for 1918, 1925, 1941, 1975, 1984, and 1994. A book was published in 1988 called 88 Reasons the Rapture is in 1988. A number of Bible scholars found firm scriptural evidence that the Rapture would happen in October of 2005. Thousands of Koreans gave away all their money and possessions in preparation for the Rapture on October 28, 1992. Even Sir Isaac Newton made a calculation based on scripture that showed the Rapture could not happen before 2060. Some Jewish scholars place the "end of days" via Armageddon in the year 2240. I couldn't find 2012 mentioned in any of these stories.

In fact, James Randi's magnum opus publication *An Encyclopedia of Claims, Frauds, and Hoaxes of the Occult* and Supernatural lists 44 distinct end of the world predictions that all came and went unfulfilled. Why should we think that the 2012 legends are any different? Any examination of the science behind any of the stories, even a glib examination, reveals a complete absence of plausible foundation. Only the Planet X story, which is the most easily falsified as it depends on concrete astronomical observations that are demonstrably false, offers a proposed mechanism for exactly how this "end of the world" is to be accomplished, the alleged gravitational destruction. Neither the Mayan calendar people, nor the Age of Aquarius people, have offered any claims for how or why the world will end, only that their particular legend points to a rollover in some ancient calendar. My calendar rolls over every time the ball drops in New York, and I've yet to see this cause any planetary cataclysm, except for the guy who has to mop out the drunk tank at the NYPD.

Many people tend to place more trust in ancient neolithic traditions than in the observations of modern science. There's nothing wrong with studying and respecting our predecessors' history for what it was, but when you turn things over and start believing that scientific knowledge of the natural world has only decreased over time, you're not doing anyone any favors.

40. Fire in the Sky: A Real UFO Abduction?

Take cover: The UFO's are coming out tonight to capture us with light beams and whisk us away to their planet for medical experiments. Today we're going to cast our skeptical eye upon the Travis Walton UFO abduction case, better known by the title of the movie made about it: *Fire in the Sky*. Among many UFO proponents, this case is considered among the most compelling, because of the number of corroborating eyewitnesses. Let's take a look, and see what happened.

In 1975, Travis Walton was a rural Arizona teenager working for his buddy (and eventual brother-in-law) Mike Rogers. Mike had a forest service contract to do odd jobs in the Apache-Sitgreaves National Forest, and this particular job was to clear brush from a 1200 acre parcel. Travis, Mike, and five buddies spent the day working, and reported the adventure of a lifetime as they drove home along a remote forest road that evening. A small silvery disk shaped UFO, about 20 feet across, came floating along. Mike stopped the truck and they watched for a few minutes. Travis thought it was pretty cool and jumped out of the truck. He ran toward it for a better view, when suddenly a blue beam of light from the UFO struck him, lifted him a few feet into the air, and while his buddies watched in terror, he was tossed like a rag doll and thrown backward into the ground on his shoulder. Mike floored it and they got the hell out of there. A few minutes later, they decided this was perhaps not the most heroic and loyal of actions, so they went back. The UFO was gone. They searched for Travis for 20 minutes, but found nothing.

Once back to town they reported the story to police, who were more than a little skeptical. Upon hearing the news, Travis' older brother Duane telephoned a UFO group in Phoenix called Ground Saucer Watch, who advised him that if Travis ever returned, to take a urine sample and bring him to

Phoenix immediately for a medical exam. After a few fruitless days of searching, Travis and Duane's mother instructed that the search be called off, which the police found a little strange.

The sheriff was not very pleased, and asked Mike and his crew to take a lie detector test. They all did, and all passed, except for one crew member whose results were inconclusive. This test was administered by an examiner named Cy Gilson, who was destined to return to the story almost 20 years later.

Five days after the abduction, Travis' brother-in-law Grant Neff said he received a midnight phone call from Travis asking him to come pick him up at a pay phone outside a gas station. Neff and Duane found Travis there, brought him home, but did not notify the police. Instead, they drove to Phoenix in the morning, to meet with the doctor promised by Ground Saucer Watch. Duane was upset to discover that the doctor, Lester Steward, turned out not to be a medical doctor at all, but a hypnotherapist.

Police were a little annoyed that they only learned of Travis' return through the mass media several days later: Neither Duane nor Mike had informed them. Still suspecting either foul play or a criminal hoax, police checked out the phone booth story. They found that the phone company did confirm the Neff home had received a call from the phone booth around midnight, but that none of the fingerprints on the phone were Travis Walton's. They found other problems too. While other people were out searching for Travis, Duane and Mike spent most of their time giving interviews to UFO investigators. Among the taped interviews that the investigators shared with the police were two interesting stories. Mike stated that he was delinquent on his forest service contract, and said he hoped Travis' disappearance would alleviate the situation. Duane said that he and Travis were lifelong UFO buffs, that they frequently saw them, and that they had recently discussed what to do if one of them were ever abducted.

There was one additional significant player in this cast of characters: The *National Enquirer* tabloid newspaper, which had a long-standing $100,000 prize offered for proof that UFOs were extraterrestrial. The *Enquirer* advised the Waltons that if they could pass a lie detector test, they might qualify for a large payment. Travis and Duane were not very keen on this idea, so the *Enquirer* agreed to keep the results secret should they not pass. The Waltons agreed. The *Enquirer* engaged an examiner named McCarthy, who, unfortunately, described Travis and Duane's results as "the plainest case of lying he had seen in 20 years." Duane was heard shouting that "he'd kill the son of a bitch." As agreed, the *Enquirer* did not publish the failed examination.

The local UFO investigators were not convinced it was a deception, however, and so they arranged a third polygraph, this time by an examiner named Pfeifer. Pfeifer reported the results as inconclusive, but the UFO group announced to the press that the results were positive and confirmed that the Waltons' story was true. This is also the examination that Travis states that he passed in his book. In later years, both of the other examiners (Gilson and McCarthy) studied the results and agreed with Pfeifer that they were inconclusive.

And that's about the point where the story fizzled out. Travis got a book deal out of it, called *The Walton Experience*, and made some money. This book is widely believed, but never proven, to have actually been ghostwritten by Jerome Clark, the editor of the *International UFO Reporter*. It's not clear whatever happened with Mike's forest service contract or whether Duane ever got any money out of the *National Enquirer*.

A lot of the information about the case, including the police suspicions and the *Enquirer's* suppressed polygraph test, was uncovered by Phillip Klass, the late full-time UFO investigator from CSICOP, now known as the Committee for Skeptical Inquiry. Apparently feeling the heat, Mike Rogers proposed a new round of polygraphs for everyone to settle the matter, under an arrangement in which if they passed, Phillip Klass would pay for the exams; and if they failed, the UFO group would pay for them. But the offer wasn't as fair as it appeared. It was only valid if Klass agreed to one particular examiner: A guy from San Diego who gave polygraph tests to plants to prove that they have feelings too.

Some 18 years later Travis' book was made into a movie called *Fire in the Sky*, which was greatly fictionalized because the studio felt Travis' own account wasn't deemed interesting enough. As part of the publicity for the movie, the studio arranged for Cy Gilson — the polygraph examiner who had originally passed Mike Rogers and the crew — to test Travis, Mike, and one of the crew again. Not surprisingly, they all passed with flying colors. But then a new face appeared on the scene, whose identity has never been known but whom Klass called simply X. Mr. X telephoned Travis and claimed to be a military intelligence operative who happened to be hunting nearby on that day in 1975. The studio had Cy Gilson test Mr. X. The only report of Mr. X's polygraph results come from the most recent edition of Travis' book, wherein he claims that Mr. X was found to be truthful about what he had seen that day, but that he was lying about being a military intelligence operative. Travis opined that Mr. X may have been hired by Phillip Klass to gain popular credibility and then publicly announce that the whole thing was a hoax, a baseless charge denied by Klass. Another possibility is that Mr. X was simply some kook looking for publicity.

So that's about the size of it. What does a skeptical analysis of the Travis Walton episode tell us? Jerome Clark, the UFO editor, has said "After more than two decades, Walton's credibility survives intact. No shred of evidence yet brought forth against it withstands skeptical scrutiny." Well, this would be true, except that there simply isn't any evidence either way. Instead, there is a gaping lack of evidence. There were no injuries to Travis' shoulder from his violent throw in the blue light beam, there were no disturbances to the pine needles on the forest floor where it all happened, and the medical exams revealed nothing to indicate any trauma or malnutrition from his missing five days. Travis and his crew have had to rely only on polygraph tests, and then only on the cherrypicked positive results, ignoring the negative results. There is just as much polygraph evidence against the Walton case as there is supporting it. This self-contradictory nature is the reason why polygraph evidence is not legally admissible in court: Speaking strictly scientifically, it doesn't tell us anything.

The few bits and pieces of physical, testable evidence that Travis' story would have produced, if true, were never present. To summarize, there is, and never has been, any proof that anything ever happened. The far more plausible explanation, that of a youthful moneymaking or attention-getting scheme by a couple of UFO enthusiasts, has worked out well. To critically analyze a far-out, incredible story like an extraterrestrial abduction, the first request we make is to show us any evidence. And, at this first hurdle, the Travis Walton story has failed completely.

41. Bend Over and Own Your Own Business

Tired of the same old grind? Want to break the mold, throw away your alarm clock, tell your boss to shove it, and become independent? Are you ready for financial freedom? Get out your checkbook, because (no matter what it is) if you want it, someone is selling it. And a lot of people are buying it. Business opportunities are made, they're not bought cheaply from ads on the Internet; nevertheless, such ads are ubiquitous. And if you lack expertise in analyzing the value of such opportunities, it's easy to get ripped off.

The basic ripoff model is a simple inversion of a standard sales rep relationship. Normally, companies selling products employ salespeople or contract with independent sales reps, who are paid a percentage of whatever orders they write. You do the work, the company pays you for your time and effort. This is the model that we're all most familiar with, and it's the way legitimate business has been conducted for centuries. Somewhere along the line, someone thought up a clever way to mix things up: A way to get salespeople to continue sending in the orders, but rather than pay them to do so, have them pay for the privilege. By re-labeling a sales job as a "business opportunity", unscrupulous companies could actually make money from the very same people they would otherwise be paying a salary to, and getting the same work for it.

Here's a typical way this works. You see an ad in the paper or on the Internet promising financial freedom, owning your own business. For some fee, say $500, you can become an authorized sales agency for XYZ Company, which sells timeshare condominiums or some other product or service. In exchange for your $500, XYZ Company will provide you with qualified leads, and you are free to pursue those leads however you see fit. Call them on the phone, knock on their door, chase them down on the street and make dramatic flying dive tackles, do whatever you can do (at your own expense, of course; you are self-employed), and hopefully get some sales. You, of course, do not have any timeshare condominiums yourself, XYZ Company does; so you need to spend a portion of the money you earned from the sale to have XYZ Company provide the product to the customer. Everything works out swell for everyone. The customer got his timeshare; you earned a profit; and XYZ Company made a sale. So what's the problem?

Well, your friend Bob was applying for a job at ABC Company at the same time you were selling your old record albums to raise the $500. Bob was given a nice office at ABC Company, was freely handed the same list of leads that XYZ Company made you pay for, and he proceeded to make phone calls on ABC Company's phone bill until he made a sale. ABC Company paid him a handsome commission, deducted nothing from it, and Bob went home for the day, secure with his employee benefits package. Bob is not only $500 richer than you, he incurred no costs of his own, and ran no risk of being poor since most salespeople like Bob are paid base salaries.

But I understand why you don't want to turn green with envy. After all, you have your freedom and are self-employed! Bob is not, Bob has to answer to his boss; and that's a lifestyle you don't want no matter how nice of a BMW Bob gets on a company lease. Your friend Red feels the way you do. Red is an independent sales rep. He sells products from various companies, and earns a nice commission on every sale. He comes and goes as he pleases, and answers to no man. But when you ask Red how much he had to pay each of his companies for the business opportunity, he looks at you like you're from Neptune. Red explains "You don't pay companies to be their sales rep, they pay you."

And now you see how you've been taken advantage of. XYZ Company has sold you on becoming their sales agent, working at your own expense and at your own risk, and also managed to take $500 from you for no good reason. If you wanted to be an independent sales agent, fine; you could easily have gone and represented any of the same companies that Red sells for, and not paid them a dime.

But don't confuse these so-called "business opportunities" with proper franchises. A real franchise, like McDonald's, is all about leveraging a proven brand. You're licensing an outrageously successful brand name, and the company provides everything you need to be a real McDonald's location. The brand name and proven business model virtually guarantees success. XYZ Company, and all the thousands of no-name companies like them, offer neither a proven business model nor a valuable brand name. They merely sell "business opportunities" to supplement their lackluster product sales, if they even have any.

You're not alone, O poor sad XYZ Company sucker. There are many, many companies out there who have figured out this simple tweak. Now they make money whether they ever sell anything or not, so long as they have an endless stream of suckers paying them to work for them. *Entrepreneur* magazine's web site provides a huge list of such "opportunities" as these:

❖ One Internet based company charges between $25,000 and $40,000 to forward leads to you for people who want their lawn mowed. You have now bought the privilege to mow those lawns, for some small commission.

❖ Another web site charges you from $399 to $50,000 for prepaid adult Internet access cards that you are then free to go out and sell. "Mom, I just started this great new business!"

❖ An ATM company charges $29,000 for the right to go out and try to sell storefronts on leasing one of their generic ATM machines.

❖ Another company charges $50,000 to $75,000, plus $2,500 annually, for the right to go door to door selling their vinyl window replacements.

❖ For an investment of only $9,800, you can become a sales rep for an Internet based travel agency, selling (you hope) the same cruise packages that normal people simply go online to buy.

- If you pay them $4,000, an office plant maintenance company will let you go out and rent plants to businesses, that you then need to service, water, and decorate for the holidays.
- Do you enjoy working with elderly people who require home care? You have two choices: Go and work for a company that provides this service, or pay $28,900 to Home Helpers for the right to go out and fend for yourself trying to find customers and bill them.

A lot of those are pretty big numbers, more than most of us have lying around handy in the bank. Thus, you'll also find that a lot of these companies offer financing. That's right: Not only do they charge you money for the right to go out and start a business that you could just as easily do on your own without them, they also make money charging you high interest on a consumer loan. You've also seen this business model on television: A lot of companies state that gold, silver, or newly issued gold coins are valuable investments, and they will even offer you a loan to purchase them. These are simply financial service companies. Their entire business is built upon these loans, which they sell to you, and then turn around and sell the loans to a major loan servicing firm. They couldn't care less what you use the loan for, whether it's buying their worthless Cook Islands gold coins or collectible plates with a picture of Dorothy and Toto. There are exceptions — but when you see that any company offers financing on their product, it's a good bet that the financing is actually their real business model, and that the product they're selling is probably not worth much. I repeat, there are certainly exceptions, but this possibility is always worth a scan of your skeptical eye. When the product they're financing is a business opportunity, you'll find that this is the case nine times out of ten.

This discussion would not be complete without a mention of multilevel marketing, or MLM. However that subject deserves an entire chapter, so I'll only mention it briefly here. MLMs are the kings of scams. You buy into an MLM by ordering say a few hundred dollars worth of product, and surveys of participants have shown that that initial purchase is all that 98% of MLM participants ever make. With this payment, you now call yourself an independent distributor, and you go forth and hope to sell those products to recoup your purchase price but also hope to sell other people to be distributors beneath you. Few people sit down to actually do the math; if even one line of distributors from one MLM program was successful in getting 5 people at each of 15 levels, this would have required the participation of four times the number of human beings than have ever existed on our planet (5^{15} = 30,517,578,125). This is not a business opportunity; this is a cleverly marketed ruse to trick people into buying a cheap product at an outrageous price, by claiming that they're buying not just a product, but also an "opportunity".

If you're considering any business opportunity, the first rule is to ignore the marketing claims presented with it and examine the deal with skepticism. If someone's trying to sell you something, that means (by definition) that the money is going into *their* pocket, not yours. Make sure that the value is genuine. People who want your services should be willing to pay you. If they're asking you to pay them, you have very good reason to be skeptical.

42. WHAT'S WRONG WITH THE SECRET?

Prepare to have everything you've ever wanted, simply by thinking happy thoughts about it; and be careful of negative scary thoughts which might cause those things to happen to you to too. Little did you know that, just like in the original *Star Trek* episode *Shore Leave*, whatever you think of — either good or bad — will actually happen! This is the premise of Rhonda Byrne's 2006 book and movie, both titled *The Secret*.

Rhonda Byrne is an Australian television producer and author. Her book and movie propose that many of the most successful people throughout history have known a "secret" — a secret closely guarded in the marketing materials for the book and movie. The "secret" turns out to be nothing more than the old motivational speaker's standby, that positive thinking leads to positive results. But she took the idea a step further. *The Secret* claims that you can actually cause events to happen by wishing for them hard enough, literally like winning the lottery or recovering from terminal illness. Similarly, a focus on fears or negative ideas will cause those things to appear or happen as well. *The Secret* calls this the "Law of Attraction". *The Secret* further makes the completely unfounded claim that many great people knew and relied upon this wisdom, and taught it to others as "secret teachers". "Secret teachers" included Buddha, Aristotle, Plato, Sir Isaac Newton, Martin Luther King Jr., Carl Jung, Henry Ford, Ralph Waldo Emerson, Thomas Edison, Albert Einstein, Winston Churchill, Andrew Carnegie, Joseph Campbell, Alexander Graham Bell, and even Beethoven. This claim is just a made-up lie: Most of these people lived before the "Law of Attraction" was invented, and there's no evidence that any of them ever heard of it.

As of today, a year and a half after its release, *The Secret* remains #26 of Amazon's list of best selling books, better than any Harry Potter book. It has over 2,000 customer reviews.

Half of them are 5 star, and a quarter of them are 1 star. This is the sign of a polarizing book. Most people either love it or find it to be utter nonsense. In the case of *The Secret*, most people love it. Thanks in large part to promotion by Oprah Winfrey, *The Secret* sold 2 million DVD's in its first year and 4 million books in its first six months.

Many of the people appearing in the movie version of *The Secret* are motivational speakers who spout the same old "If you can dream it, you can do it" nonsense that Amway salesmen have been chanting for decades. In essence, part of what Rhonda Byrne has done has been to simply repackage Motivational Speaking 101 inside the wrapper of a century-old philosophical construct, which we'll look at in closer detail in a moment.

As you've probably heard, *The Secret* has been roundly criticized from all quarters. The most common criticism is of *The Secret's* assertion that victims are always to blame for whatever happens to them. Whether it's a rape victim, a tsunami victim, or a heart attack victim, *The Secret* teaches that they brought it upon themselves with their own negative thoughts. This idea is, of course, profoundly offensive in many ways. Doctors attack *The Secret* for teaching that positive thinking is an adequate substitute for medical care in cases of serious illness: Wish for it hard enough, and your cancer tumors will melt away. Religious leaders criticize *The Secret* for its ethical claims that victims are always to blame, and for promoting the attitude that anyone can be just like a god by wishing hard enough. Many financial critics and advisors have pointed out the dangers of yet another baseless get-rich-quick scheme. The list of critics of *The Secret* goes on and on, as tends to happen to any mega-successful franchise.

So the question people ask me is "What do I think of *The Secret?*" This is really asking what is the best way to use critical thinking to analyze the validity of *The Secret's* claims. To do this, we first ignore everything that people say about it. We ignore the critics, we ignore the supporters and testimonial writers, and we ignore the Amazon reviews. Let's examine the claims themselves, on their own merits, and let's start by tracking down precisely where this "secret" of the "Law of Attraction" comes from.

The concept now called the "Law of Attraction" was described by James Allen in his 1902 book *As a Man Thinketh*. He wrote: "The soul attracts that which it secretly harbors, that which it loves, and also that which it fears. It reaches the height of its cherished aspirations. It falls to the level of its unchastened desires — and circumstances are the means by which the soul receives its own." Allen was saying that circumstances — things that happen to us — will make our desires and our fears both come to pass. Allen said that our desires and fears would "attract" those things. If Winston Churchill was indeed a "Secret teacher", we might conclude that he desired gin and feared the fire bombing of London, because both of those things certainly found their way to him. Allen wrote his book during a philosophical period called the New Thought movement, which applied metaphysical concepts to modern life. This movement was akin to what we describe as New Age today: Same ideas, slightly different buzzwords, a century apart.

Other authors followed suit based on James Allen's success, and the term "Law of Attraction" came into being among some of these followup books. A hundred years later, Rhonda Byrne read Wallace Wattles' 1910 New Thought book *The Science of Getting Rich*, and cleverly used it as an "ancient wisdom" foundation for contemporary motivational self-help ideas. The general public tends to love anything that can be attributed to ancient wisdom, so it's no accident that Rhonda made reference to Buddha, Aristotle, and Plato.

New Thought's "blame the victim" concept is one that's attractive to most people at a deep level. When we see someone else victimized, we take a sort of smug pride in that we did not let that happen to ourselves because we did not think whatever ugly thoughts that person must have. *The Secret* works! *The Secret* appeals to that selfish ego that's somewhere inside of all of us. This is ugly and embarrassing, but it's part of why *The Secret* is psychologically appealing.

Put all of these together, and The Secret is a marketing 1-2-3 punch:

- ❖ It's based on ancient wisdom, which is always popular.
- ❖ It sells the same motivational self-help pitches that are always popular.
- ❖ It teaches that you're already a winner because you didn't fail like those people who died in New Orleans.

Some claims in *The Secret* are simply factually wrong, and so fall apart under their own weight when scrutinized. Specifically, *The Secret* claims that quantum physics supports and explains the "Law of Attraction". At its most superficial, this claim sounds reasonable to the uncritical layperson because attraction sounds like magnetism which is a real scientific thing, and any mention of the term quantum physics sounds scientific enough to be acceptable at face value. Who's qualified to argue against quantum physics? *The Secret* says that thoughts have energy, and similar energies are attracted to each other. That's their quantum physics.

In fact, scientifically speaking, that statement is completely meaningless at every level, and at no level does it have anything whatsoever to do with real quantum physics. In fact, the closest analog I can find in science is that like charges repel one another, they do not attract. But we're talking about "thought energy" here, so we've already left the realm of real science and are in the world of metaphysics. Since metaphysics is a philosophical construct with no connection to real physics, either quantum or classical, you can pretty much say whatever you want and there is no scientific way to respond to it. Thus, *The Secret's* claim to have roots in quantum physics is childish and meaningless, yet it succeeds because it appeals to the uncritical layperson's tendency to accept scientific sounding terminology at face value. Check out Rhonda Byrne's background in quantum physics. You'll find that she took the same university courses that your cat did.

Now, it's probably important to point out that there's nothing wrong with positive thinking, and usually nothing terribly helpful about negative thinking. People with positive attitudes tend to be happier and more personable. People with negative attitudes tend to bring other people down or get blown off. In this sense, having a positive attitude is good, but nobody needs to be told that and you certainly don't need a self-help book and movie to make the point. The important line to be aware of is the division between fantasy and reality. People who buy into *The Secret* are not generally healthier or wealthier than anybody else, in fact they're poorer by the price of a movie ticket or a book. So go forth and be a positive person, but of claims that thought materializes into physical possessions or actions, you have good reason to be skeptical.

43. THE FACE ON MARS REVEALED

Imagine yourself in a NASA control room, late at night. The coffee's cold and, outside, the rain drums steadily against the window. You start to drowse off in your chair, when suddenly the teletype jumps to life with a loud mechanical bang. You're startled, but annoyed; and as it starts hammering out its latest data, you try to go back to sleep. You've heard this all before and seen a million badly printed images. But then, as it finishes printing the second page, your eye catches that long sheet of perforated printer paper folding into a pile on the floor, and you see something unbelievable. There, in yet another series of photographs from Mars, is a distinct human face.

It was 1976, and Viking I was sending its latest images. Among a number of similar hillocks and mesas in a region of Mars called Cydonia Mensae, one feature stood out. It was a clear rendering of a human face! NASA engineers loved it; they passed it around, put it out for publication, and had all sorts of fun with it. But what they hadn't anticipated was that some in the public thought it was actually an artificially carved human face, despite the accompanying explanation that it was just a hill that happened to have this funny resemblance to a face when the light was at a certain angle. One of it most important distinguishing features, a nostril, was only one of many black dots that actually represent missing data in the image. Before long, to the dismay of astronomers worldwide, there was a firmly established pop-culture belief that there was a real gigantic human face on Mars, carved in perfect detail by aliens.

As the decades wore on, better cameras took better images, most recently culminating in the 2007 image taken by the HiRISE camera aboard the Mars Reconnaissance Orbiter, with a super high resolution of about .3 meters per pixel. The Cydonia "face" turns out to be merely an unremarkable hill, with plenty of natural random variations on its surface, and no longer looks anything remotely like a face or any other kind of carving. However you can see the general contours that made up the facial features in the original image. While those black dots of missing data in the original image gave the illusion of sharp focus, the image is now shown to have been extremely blurry. Although a two-dimensional view of the hill does have the appearance of some symmetry, the improved image shows that it's nowhere near as symmetric as it appeared to be in the original blurry image.

The popular belief in an artificial sculpture would probably have never emerged if not for the writings of conspiracy theorist Richard Hoagland. Hoagland saw the original image, immediately concluded that an artificial carving was the only reasonable explanation, and wrote the book *Monuments of Mars* claiming that the Cydonia face is only one of many artificial structures on Mars, including pyramids and whole cities. He claims that NASA has exhaustive photographic evidence of all such structures, but that they cover them up and suppress them to avoid the mass panic that would inevitably ensue should Hoagland's claims be proven. Other claims of Hoagland's include taking credit for designing the plaque that was on Pioneer 10, which was done by Carl Sagan and which Hoagland had nothing to do with; that he first conceived the idea of subsurface oceans on Europa in a 1980 article, even though scientists including Isaac Asimov had been proposing this throughout the 1970's; and that a concept involving trans-dimensional energy that he calls "hyperdimensional physics" is correct and that every educated and professional scientist is wrong about the nature of the universe. Since he has positioned himself as the leading public advocate for NASA's evil coverups and the "truth" about Martian civilization, I recommend that you at least read through the Wikipedia entry about Richard Hoagland to see just how disconnected this guy really is. And then, you might want to reconsider your sources if you take seriously any claims that Hoagland has promoted. Of course, this ad-hominem attack against Hoagland personally says nothing directly about the validity of his claim that the face is an artificial structure, so I'll refer you to the high resolution photos to make up your mind about that for yourself.

Geological features that happen to look like faces, people, or other objects are not rare. In Alberta Canada, there's a figure called the Badlands Guardian that, when viewed from the air, looks astonishingly like a Native American wearing a full headdress and listening to an iPod. In fact, it looks way more like a person than the Cydonia face ever did on its best day. Why hasn't Richard Hoagland claimed that somebody carved the Badlands Guardian? That would be a lot more plausible. He probably doesn't make that claim because it would be testable and easily falsified.

But the Badlands Guardian is only one example. The Old Man of the Mountain in New Hampshire looked just like the profile of a man jutting out from a cliff until it collapsed in 2003. North Carolina has a giant head sitting on a cliffside ledge called the Devil's Head. Sundance, Wyoming is home to the Old Man of the Park, and the Absaroka Range in Montana features an amazingly lifelike face called the Sleeping Giant near Livingston. But, these are all newcomers. From the day the first protohuman looked into the sky, we have marveled at the Man in the Moon, the largest facelike structure known.

Although some of these features are pretty astonishingly realistic, they don't even have to be. Your brain will still say "Face", even if it's as indistinct as the Cydonia face. This is a perceptual phenomenon called pareidolia, or apophenia, which is the tendency for the brain to see order in randomness. The famous Rorschach inkblot test is based on pareidolia. Pareidolia causes cryptozoologists to see crouched Bigfoots in forest photographs. It causes us to see a face made of headlights and grills on the front of a train or a truck, or the face of the Virgin Mary on a piece of toast. Pareidolia means that any two dots and a line, like those on the Cydonia face, will shout "eyes and mouth" to a human brain. Carl Sagan proposed that brains are hardwired to see faces. Without the phenomenon of pareidolia, no drawing less than a Rembrandt masterpiece would be recognizable as a face.

But let's even set pareidolia aside, and just look at the original blurry photograph of the Cydonia face. The face is about one square kilometer. The entire surface of Mars is about

150 million square kilometers. Thus, if we postulate that the Cydonia face is about a one-in-a-million oddity, probability dictates that somewhere on the surface of Mars, some 150 one-kilometer areas bear some equally improbable likeness. How many fist-sized rocks are there in a square kilometer of Martian surface? A million, maybe? If one in a million fist sized rocks bears some resemblance to J. Edgar Hoover, we should expect to find 150 million fist-sized rocks on the surface on Mars that look like J. Edgar Hoover. By the sheer weight of large numbers, it's a virtual certainty that the surface of Mars has natural structures that look like human faces, elephants, and Ferraris, when viewed from certain angles. By the same law, you'll also find these things on Venus, Titan, Pluto, and Halley's Comet.

So let's collect all of our evidence about the Cydonia face:

❖ It doesn't actually look anything like a face.
❖ Pareidolia gives us pretty loose parameters to decide what qualifies as a face.
❖ Probability absolutely requires hundreds of startlingly good faces on Mars, and on any other planet.
❖ We've never found any evidence of sculptor civilizations on Mars.

So, are the conspiracy theorists right, that an artificial sculpture is the most likely explanation for the Cydonia face? Well, I have to conclude that the evidence for that is pretty thin. A much better alternate explanation is available: That it is a simple natural structure, which when viewed with the details blurred out and with right lighting conditions, can look like a face, much like countless natural structures on Earth. What do you think?

44. The Crystal Skull: Mystical, or Modern?

It was 1926 when Anna Mitchell-Hedges, adoptive daughter of British adventurer and author Frederick Albert Mitchell-Hedges, was something of a real life Lara Croft. She was crawling through an ancient Mayan temple in Belize, long ago wrecked by the ages and the ravages of the encroaching jungle. Beneath a crumbled altar, she unearthed perhaps the most curious artifact from the ancient world: A perfectly clear crystal skull, expertly carved, and immaculately preserved, and about two thirds the size of a real skull. For nearly 30 years the Mitchell-Hedges family kept the crystal skull a secret, until F.A. Mitchell-Hedges mentioned it briefly in his book *Danger My Ally*. In this book he said the skull was 3,600 years old, and was used by Mayan priests to strike people dead by the force of their own will. After her father's death, Anna took this so-called "Skull of Doom" on tour throughout the world, and its strange powers became well known. Arthur C. Clarke even used the Mitchell-Hedges skull as the logo for his television series *Arthur C. Clarke's Mysterious World*. The fourth *Indiana Jones* movie is about a crystal skull with mystical qualities, and furthers the theme originally proposed by Mitchell-Hedges that crystal skulls are alien in origin, coming from Atlantis or Roswell or some alien world. In fact, practically every reference to a crystal skull over the past 40 years or so has usually been specifically about the Mitchell-Hedges skull.

Some believers in mystical energy feel that the crystal skulls have a broad range of powers. They can be used to aid in divination, in healing, and even psychic communication. Others claim that they have refractive properties unlike other crystals. They are said to remain at exactly 70 degrees no matter what temperature they are exposed to. They possess spiritual auras that can be photographed. Some even speculate that when all the crystal skulls are brought together, it will bring about the end of the world.

Now, I'm reluctant to burst anyone's bubble, but before going further it's necessary to clear up a few misconceptions. The Mitchell-Hedges skull is not quite 3,600 years old, and Mitchell-Hedges found it a little closer to home than Belize. In fact, he bought it from Sydney Burney, a London art dealer, through a Sotheby's auction on October 15, 1943, as determined in hard black and white by investigator Joe Nickell and others. This explains why neither Mitchell-Hedges nor his daughter ever said anything about it following their alleged 1926 discovery. They had never heard of it, until they bought it 18 years later, and then invented their Mayan altar story.

So this Sydney Burney character, perhaps he was the one who actually found the skull in a Mayan ruin, and traced its history back to the Atlanteans? Well, there is additional hard evidence that Burney owned the skull as far back as 1933, because he wrote a letter about it to the American Museum of Natural History, which they still have. Three years later, the British anthropological journal *Man* published an article about Burney's skull.

It seems clear, but has never been never proven, that Burney bought the skull from French collector Eugene Boban. The timing was right; the two men knew each other; and Boban is known to have sold at least two other crystal skulls about the same time Burney acquired his. If Mitchell-Hedges was the real Indiana Jones, Eugene Boban was the real Belloq. He was even French. And, like Belloq, he didn't actually go into the jungle tombs personally to acquire his artifacts. In Boban's case, he simple purchased them in bulk from the manufacturer. This time, the manufacturer was Germany's so-called "capital of the gemstone industry" Idar and Oberstein, a bucolic hamlet where artisans and craftsmen chip away at semi-precious stones in their workshops like so many Gepettos. In the 1870's, craftsmen in Idar and Oberstein made a large purchase of quartz crystals from Brazil, from which to make carvings. Nobody has ever found documented proof, but at about the time the Idar and Oberstein craftsmen were selling their cunningly carved art objects of Brazilian quartz, Eugene Boban left from there with at least three, and possibly as many as thirteen, freshly carved skulls made from Brazilian quartz. Any connection you choose to draw is purely speculative. According to documents found by Jane Walsh, a Smithsonian archivist, Boban sold one of his skulls to Tiffany's in New York City, which in turn sold it to the British Museum in 1897. Boban sold a second skull to a collector who then donated it to the Museum of Man in Paris. An 1887 *New York Times* article describes the British Museum skull then in the hands of New York collector George H. Sisson, who had bought it from Eugene Boban. After this article, Sisson sold this skull to Tiffany's.

For decades, the British Museum and the Museum of Man displayed their crystal skulls with the provenances originally provided by Eugene Boban, which was that the skulls came from pre-Columbian Aztec origin. But then, in separate studies in the 1990's, both the British Museum and the Smithsonian examined a number of crystal skulls, including all of those in museum collections attributed to Eugene Boban. Analysis of the cut and polish marks by electron microscope proved that they were made using 19th century rotary cutting tools, identical to those in use in Idar and Oberstein at that time. The British Museum now lists their skull as "probably European, 19th century," and "not an authentic pre-Columbian artifact."

The Paris skull, also from Boban, was subjected to even better tests in 2008, confirming that its polishing was done using modern tools. In addition, particle accelerator tests found traces of water used during the cutting and polishing, occluded within the quartz, that positively dated the carving to between 1867 and 1886.

Neither the Mitchell-Hedges nor their skull's current owner, family friend Bill Homann, ever allowed the Mitchell-Hedges skull to be tested with modern equipment; nor have any of the owners of other famous crystal skulls like the one called Max in Texas. The privately owned skulls now confine themselves to touring to mysticism conventions, New Age hotbeds like Sedona, and charging for private viewings and sessions. So far as I've been able to find, no private crystal skull owner has ever allowed controlled tests of their claims of any mystical powers they say their skull has. If they'd like to, this is my personal guarantee to fast-track them to the James Randi Educational Foundation's million dollar prize.

There is enough of a gap in the early history of the Mitchell-Hedges skull that we cannot absolutely trace its lineage from the Idar and Oberstein workshops in the 1870's to the hands of Sydney Burney in 1933. Everything known about the skull is consistent with that history, and no evidence has ever been presented that the skull might have any other origin. There is the Mitchell-Hedges' own story of having found the skull in their pulp-fiction Mayan tomb adventure, but that story has been conclusively proven to be a fabrication by documentation from Sotheby's and Burney.

All of this makes it rather difficult to form an opinion about the mystical powers of crystal skulls. If these powers are attributed to their Mayan, Atlantean, or alien origin, then that attribution is conclusively false, but that doesn't mean the mystical power itself doesn't exist. The first thing the claimants would need to do is articulate exactly what the supernatural power is, and then demonstrate it under controlled conditions. Neither of these has ever been done, so a truly critical analysis has nothing to advance it beyond a null hypothesis. And so there we have it: All known crystal skulls are of modern origin, with no unusual properties, and no coherent or testable claims of anything out of the ordinary. *Indiana Jones* might make great entertainment, but it makes poor archaeological history.

45. Reassembling TWA Flight 800

July 17, 1996 — My wife and I were on our honeymoon, flying out of New York City. We got out all right, but one of the planes just behind us wasn't quite so lucky. TWA Flight 800, an older Boeing 747 jumbo, took off and headed out over the Atlantic Ocean. About twelve minutes after its departure, at about 13,700 feet, an explosion broke the aircraft in half just forward of the wing. All 230 people on board were killed.

The NTSB (National Transportation and Safety Board) managed to recover all 230 bodies, and over 95% of the wreckage from the ocean floor, which is pretty incredible. They reconstructed the aircraft to understand what went wrong. What was found, and what nobody disputes, is that the principal destruction was caused by an explosion of the fuel in the center wing fuel tank. What has never been determined is what triggered that explosion. Conspiracy theorists immediately jumped on this, and concluded that the aircraft must have been shot down by the US government, either deliberately or as accidental friendly fire.

Fuel was thrown on this fire from two principal sources. The first source was a number of eyewitness reports from people who saw a second brightly lit object going up into the sky and contacting the aircraft, a description which certainly sounds like a missile attacking the aircraft. The second cause of conspiratorial speculation was triggered by the government itself. The FBI, who was investigating the crash to see if it was a criminal act, engaged the CIA to produce a computer animation showing what the aircraft did after it exploded, in order to answer the questions of the families of the victims. According to the NTSB, when the nose broke off the aircraft, that made it tail heavy and it veered sharply upward for several thousand feet, burning all the way, thus looking like a missile. The FBI had no available computer animation resources of their own, so they had the CIA do it. And, once the CIA became involved, that screamed out to every conspiracy theorist in the world that the whole operation was a clandestine government coverup.

The aircraft was two and a half miles up, and about nine miles offshore, when it exploded. That puts the coastline just about exactly one minute away at the speed of sound. The vast majority of the eyewitnesses were between one minute and two minutes away, as sound travels. The majority of the 38 eyewitnesses who reported a skybound streak that's been described as a missile trail only turned to look after they heard the explosion. This means that for at least two minutes after the plane exploded, something happened that looked to many eyewitnesses like a missile going up. Remember, the majority of people who reported that it looked like a missile struck the aircraft, did not start watching until at least one minute after the explosion happened. Therefore, in most cases of people who said it was absolutely a missile, the laws of physics make it impossible that they could have seen such a missile. We know for a fact that what the aircraft did one minute after it exploded, looked enough like a missile to convince many eyewitnesses that it couldn't possibly have been anything else. In all of these cases, whatever they saw happened after any theorized missile would have detonated.

One of the conspiracy web sites, Flight800.org, has a page giving testimony from witnesses who believe that they distinctly saw two separate objects, a missile and a plane, converge. As you read, pay attention to *when* the witnesses heard the sound relative to what they saw:

Witness 73
> ...While keeping her eyes on the aircraft, she observed a 'red streak' moving up from the ground toward the aircraft at an approximately a 45 degree angle. The 'red streak' was leaving a light gray colored smoke trail... At the instant the smoke trail ended at the aircraft's right wing, she heard a loud sharp noise which sounded like a firecracker had just exploded at her feet. She then observed a fire at the aircraft followed by one or two secondary explosions which had a

deeper sound. She then observed the front of the aircraft separate from the back.

Witness 88:
...All of a sudden he heard an explosion. He glanced over to the southeast and observed what he thought was a firework ascending into the sky. All of a sudden, it apparently reached the top of its flight... At this point he observed an airplane come into the field of view. He stated that the bright red object ran into the airplane and upon doing so both the plane and the object turned a real bright red then exploded into a huge plume of flame.

Witness 675:
...Noticed an orange flare ascending from the south... trailing white or light gray smoke. He then observed the flare strike what looked like an eastbound Cessna airplane on the port side... Within five (5) seconds... he heard what sounded like thunder and felt the ground shake.

Witness 145:
...She saw a plane and noticed an object spiraling towards the plane. The object which she saw for about one second, had a glow at the end of it and a gray/white smoke trail... She heard a loud noise and saw an explosion just as the object hit the plane. The plane dropped towards the water and appeared to split in two pieces. A few seconds later, she heard another explosion.

Whether you're a conspiracy theorist or not, the 1-minute minimum delay required by the speed of sound clearly makes it impossible to corroborate what these people heard with what they think they saw. This illustrates why the witness testimony, while still valuable, cannot be relied on as the definitive explanation for what happened. Anecdotal evidence has value for suggesting directions to research, but it does not by itself constitute evidence, and cannot reasonably be treated as such.

Anyway, who *could* have fired a missile? The FBI did identify some military assets that were in the area at the time, including a US Navy P3 Orion aircraft, and a US Coast Guard cutter. Neither asset has an anti-aircraft or missile capability. Radar data from four different sites also found four unidentified boats within six nautical miles of Flight 800, all but one of which responded to assist in search and rescue. Shoulder launched weapons do not have anything like the range required to reach the aircraft from the shore, or the ability to reach the plane's altitude if fired from the water.

There's one final loose end that nobody has been able to definitively tie up, and that's the discovery of explosives residue on the debris. Although the conspiracy theorists charge that the NTSB has covered up this discovery, in fact the NTSB has freely and openly disclosed everything about it. It's known that no high energy explosives detonated on board the aircraft, because there is zero evidence of explosives damage anywhere. The best theory is that this residue is left over from exercises conducted with bomb-sniffing dogs on board the plane several months before. Conspiracy theorists charge that no such tests were conducted aboard this plane, but all available records indicate that they were. As a result of this theory, the NTSB made recommendations changing the procedures of such tests to prevent such explosives residue from contaminating other aircraft in the future.

If you do a Google search for "TWA Flight 800", most of the results are from conspiracy web sites that uncritically start with the assumption that the US government shot down the aircraft. These web sites then present opinion, conjecture, and hypothetical extrapolation that support that assumption. Sometimes you'll hear conspiracy theorists charge that the NTSB ignores eyewitness reports, or suppresses anything that doesn't agree with their official story of an accident. Anyone who's a pilot or an aviation nut knows that this couldn't be further from the truth. Go to NTSB.gov and click on Aviation Accident Database. Search for some recent accidents, as these will show you what an investigation looks like in progress. What you'll see are the facts that are known, and you'll see any eyewitness reports there might be. What you won't see is anything like an explanation or a theory that tries to explain the evidence, and certainly nothing like an "official story" that anyone is sticking to.

If you want to see what a final report looks like, go to NTSB.gov, click on Accident Reports, and click on Older Aviation Accidents. You'll see TWA Flight 800 near the top of the list. The final report is a huge 341-page PDF document, but you can also just read the four paragraph summary. Here's a quote from it:

> *The source of ignition energy for the explosion could not be determined with certainty, but, of the sources evaluated by the investigation, the most likely was a short circuit outside of the CWT (center wing fuel tank) that allowed excessive voltage to enter it through electrical wiring associated with the fuel quantity indication system.*

In the full report, the NTSB goes through other possible causes for the explosion of the center fuel tank, including:

> *A lightning or meteorite strike; a missile fragment; a small explosive charge placed on the CWT; auto ignition or hot surface ignition, resulting from elevated temperatures produced by sources external to the CWT; a fire migrating to the CWT from another fuel tank via the vent (stringer) system; an uncontained engine failure or a turbine burst in the air conditioning packs beneath the CWT; a malfunctioning CWT jettison/override pump; a malfunctioning CWT scavenge pump; and static electricity.*

There's no need to repeat their findings on each of these causes here, if you're interested you can grab the full report and read section 2.3.1. In each case, the potential cause was found to be unlikely, unsupported by any evidence, and lacking evidence that would have resulted. Section 3.1 of the report lists their findings, which are facts that were determined with certainty. Among these: "The in-flight breakup of TWA flight 800 was not initiated by a bomb or a missile strike."

Of course, this doesn't change the mind of a die-hard conspiracy theorist, because this government-produced paper is simply part of the conspiracy. In fact, they consider the report's very existence as further evidence of the conspiracy. When you hear a conspiracy theory that provides no testable evidence of its own, but relies only on anecdotal testimonies, extrapolations of possible motivations, and non-evidenced claims of implausible coverups, you have every good reason to be skeptical.

46. Is Peak Oil the End of Civilization?

Think back to the days following 9/11, when all commercial and civilian air traffic was grounded. People were stranded everywhere. Only take it a step further: Take away the trains, the buses, and the rental cars. Imagine every gas station closed, and cars abandoned on the roadside as they too run out. People can't get to work and the nation's businesses can't even declare bankruptcy because there's no way to get to the courthouse to file the papers. The mail stops. Supermarkets are empty because there are no delivery trucks. And then, in a final shriek of terror, the power plants shut down, darkness falls everywhere, water pressure stops, and humanity devolves into a battlefield of hand-to-hand combat over who gets to eat the neighbor's dog.

This is one extremist scenario painted by peak oil advocates. Peak oil refers to the point at which world oil production must start to decline as reserves are emptied and pumps run dry. Oil is a finite resource, and so there's no doubt that at some point, peak oil will occur. The world's appetite for oil continues to grow exponentially, fueled by the explosive growth of most of the world's population in China and India. When the rift between increasing demand and decreasing supply gets to a breaking point, advocates say that the apocalyptic scenario described above must happen.

In some places, peak oil has already happened. In the United States, oil production peaked about 1970, when we produced about 3.4 billion barrels per year. Today we produce about 1.5 billion. The curve has followed the 1956 prediction by American geophysicist M. King Hubbert who described the oil production of any given region over time as a bell curve. This is called Hubbert's curve. Every region in the world has its own separate curve. Some, like the US, are already on the downside. Others, like Canada, which is just beginning to exploit its oil sands, are only just now hitting the steepest climb on the upside. In addition to Canada, the middle east and China are also still climbing their upsides. Russia has just barely tipped past its peak. The most pessimistic estimates for world peak oil say that it's already happened; the most optimistic give us another 30 years before we peak.

Oil production in any given region is not determined simply by physical factors such as the amount of reserves remaining and the technology required to develop it, but also by political and economic pressures. For example, Russia peaked way back in the Soviet days when their economy was falling apart, but their oil industry recovered throughout the 1990's and they've managed to find a second peak. The same could happen in the United States, if continental shelf and rocky mountain reserves were to be developed. They probably won't be, due to political pressures, but it's nice to know that they could be, if things came down to eating your neighbor's dog.

The biggest error made by the peak oil doomsayers is in failing to recognize the adaptive nature of the world economy. When demand goes up and supply goes down, prices go up, and consumers look to alternatives. As alternatives become more popular than the original, prices drop in reaction to the reduced demand, and eventually a marginalized industry disappears. Markets react and adapt. Currently, we have high gasoline prices. Consumers are reacting by clamoring for alternative fuel vehicles. Many industrial products depend on oil, such as fertilizers, solvents, and plastics to name only a few; and as the price of producing these climbs, industry turns to alternatives. Alternatives become increasingly commoditized and prices come down. Oil becomes less relevant, and eventually nobody will care when reserves finally do run dry.

If consumers and industry failed to react to oil prices that climb astronomically for decades, then yes, it could be possible that we'd have an overnight shutdown of everything and the world would turn into a tumultuous battlefield of cannibalism. But neither consumers nor industry have ever acted this way in the past, don't now, and aren't likely to in the future. Everyone wants to spend less money, and the most expensive options will always be the least desirable.

Part of the reason that the doomsayers don't see this is that what's most visible right now is gasoline prices, and the total nonexistence of any viable alternative fuel vehicles. That's our worldview: Expensive gas, diminishing production, no alternatives. But outside of this worldview, some very interesting things are happening. Believe it or not, tremendous research is going on to develop alternative fuel vehicles: Supercapacitor technologies, fuel cells, hydrogen production. Silicon Valley is investing into alternative energy like a bat out of hell. And worldwide, industry is developing alternatives to petroleum based plastic like fructose. Agricultural fertilizers can be made from seawater or atmospheric nitrogen, but some of these options are not yet in wide production because the market has not yet reacted that far. Eventually, when the price gap closes, these non-oil based sources may become the inexpensive standards.

What about other resources? Do they peak as well? Yes, they do, at least non-renewable ones do. It's generally believed that we're either just past or right about at the peak of gold production. Some 150,000 tons of gold have been mined throughout history, and the US Geological Survey estimates that there's another 90,000 tons still out there. Considering there's been a constant increase in gold mining efficiency, this all sounds about right. In many countries around the world, gold production has been dropping in recent years as reserves have been tapped out. But, do we need to expect a worldwide panic over gold? Probably not, since it's largely a luxury item and its industrial uses are relatively modest. We expect to see prices rise as supply diminishes, and probably a number of market adjustments until we settle into an eventual equilibrium of old gold being reused to meet demand.

There are other more serious resource peaks. Peak phosphorous, for one. Phosphorous is a crucial ingredient of both synthetic and organic crop fertilizers. And glyophosphate is a principal ingredient in herbicides used in super-high-yield genetically modified food crops. Both have seen dramatic price hikes in recent years as rock phosphate, the source of nearly all industrial phosphorus, is being mined out. If I stopped talking now, this would seem an alarming, terrifying prospect, and you might see doom & gloom web sites predicting global disaster the way you do with peak oil. Peak phosphorus is a painful situation for farmers, but it's one that's not insoluble. In the short term, farmers invest in the phosphorus companies, thus offsetting their production costs with dividend income. In the long term, the fertilizer producers continue their research into alternative supplies. High on their list is seawater, which is after all, the eventual depository of all agricultural runoff. This proposal is essentially an accelerated leveraging of nature's existing cycle.

Another peak that we're starting to hear about is peak silicon. In this case, it's not a physical shortage of silicon; it's the engineering limits of what you can do with silicon to make computer chips. Such a peak would mean that the capabilities of computers could no longer grow with our increasing demand on them. The doomsayer pundits could make an argument here too that overnight we'll start each eating other and burning down our cities and running around with babies on pitchforks. But in fact what happens is that the silicon industry fades and the graphene industry rises. Graphene is only one of numerous next-generation computer chip technologies that obsolesce silicon.

Generally, what tends to happen in any industry, is that by the time an existing resource runs out, inventive scientists have already come up with something better. When a production peak looms (be it oil, phosphorus, silicon, or anything), this provides a kick in the pants to accelerate development. Market economies work in such a way that investors are encouraged to fund such development, and the bigger the looming problem, the bigger the investment to meet it.

Mark Twain used to speak nostalgically of the sad disappearance of the riverboat industry. It was killed by steam trains. Steam trains were later killed in turn by diesel electrics. The glory days of rail travel were ended by the advent of airlines. Eventually airliners aren't going to be able to burn jet fuel anymore. What will happen then? Today, I have no idea; much like Mark Twain had no idea that airliners would one day replace his beloved steamboat. To assume that the current state of technology represents the last and final stage of development is a completely ignorant viewpoint. To conclude that the doomsayers must be right is to be completely uncritical and unskeptical. I don't know what's around tomorrow's corner, but all the evidence of history tells us that it's probably not a big scary dragon.

47. WHAT YOU DIDN'T KNOW ABOUT THE STANFORD PRISON EXPERIMENT

It was 1971 when the prisoner, emotionally drained, sleep deprived, chained, and dehumanized in his rough muslin smock was thrown into a tiny dark closet by the cruel guard nicknamed John Wayne, to endure solitary confinement without food or bathroom privileges. You might think this scene was from Hanoi in Vietnam, or at best a military prison in the United States. You'd be close. This brutal activity was funded by the United States Navy, which was interested in learning more about the psychological mechanisms in a prison environment. It took place at Stanford University in California, and the prisoner had done nothing wrong other than to volunteer for a research project. This was the infamous Stanford Prison Experiment, conducted by professor of psychology Dr. Philip Zimbardo.

Philip Zimbardo grew up in what he describes as a "South Bronx ghetto", and as a boy watched his close friends engage in acts of violence, abuse drugs, and wind up in jail. He grew fascinated by the question of why good people do bad things, and became convinced from a very young age that bad environments tend to poison the people placed into them. Put a good person into an evil situation, and that person will become evil. He later wrote:

> To investigate this I created an experiment. We took women students at New York University and made them anonymous. We put them in hoods, put them in the dark, took away their names, gave them numbers, and put them in small groups. And sure enough, within half an hour those sweet women were giving painful electric shocks to other women within an experimental setting... Any situation that makes you anonymous and gives permission for aggression

247

will bring out the beast in most people. That was the start of my interest in showing how easy it is to get good people to do things they say they would never do.

From his body of work, it is easy to conclude that he was actively interested in justifying a preconceived notion: That good people will become evil if you put them into an evil environment. About a decade after getting his Ph.D. in psychology from Yale, Zimbardo went to Stanford University, where he got tenure and then set about planning the experiment that was to define his career.

24 students were recruited for a two-week experiment for which they would each receive $15 per day. They were randomly assigned to be either prison guards or inmates. The prisoners were surprised to be picked up unexpectedly at their homes by real Palo Alto police officers. They were roughly hustled to their new home, stripped, deloused, and put into rough muslin smocks with no underwear. Zimbardo described it:

> *The question there was, what happens when you put good people in an evil place? We put good, ordinary college students in a very realistic, prison-like setting in the basement of the psychology department at Stanford. We dehumanized the prisoners, gave them numbers, and took away their identity. We also deindividuated the guards, calling them Mr. Correctional Officer, putting them in khaki uniforms, and giving them silver reflecting sunglasses like in the movie Cool Hand Luke. Essentially, we translated the anonymity of Lord of the Flies into a setting where we could observe exactly what happened from moment to moment.*

The results have become legendary. Some of the guards seemed to relish their newfound authority a little too much, becoming sadistic, and working extra hours just for fun. The torment they put on the prisoners was real. Some began showing physical manifestations of stress and psychological trauma, to the point that one third of them had to be removed from the experiment early. In fact, it got so bad that Zimbardo decided to end the experiment after only six days, less than half the planned duration.

Zimbardo's conclusion was clear. Good, ordinary college students willingly became sadistic tormentors, simply because they were given the permission, the means, and the expectation of doing so. The Stanford Prison Experiment, and this well-publicized result, became a permanent fixture in the popular conception of psychology.

The problem is that a lot of the psychology community disagrees with his findings. Some found that any results were rendered meaningless by insufficient controls. Some have problems with his analysis of the results, reaching a different conclusion based on the same data. Some found the sample population invalidated by selection biases, or the size of the sample inadequate for statistically useful results. Some found methodological flaws that tainted the participants' behavior. Let's look at some of these criticisms in closer detail.

❖ First, the issue of selection bias. Selection bias is where you choose your subjects in such a way that they are not truly representative of the general population. In this case, Zimbardo advertised to students to participate in an experiment about "prison life". Clearly, a large segment of the general population would be repulsed by such a concept, and you've got to have questions about anyone attracted to that idea. Thus, all applicants to the Stanford Prison Experiment were preselected for comfort with the idea of "prison life".

❖ Most of the Stanford guards did not exhibit any cruel or unusual behavior, often being friendly and doing favors for the prisoners. The most notorious guard, nicknamed John Wayne, explained that he was simply trying to emulate Strother Martin's character from Cool Hand Luke. Other analysts have found it difficult to support Zimbardo's conclusions, since the allegedly poisonous environment did not affect most participants, and the most notorious participant explained that his motivation came from a completely different source.

❖ Zimbardo himself was also criticized for actively participating in the experiment as one of the characters. He was the prison superintendent. Although he may have restrained himself from having any influence on the experiment, the fact that he put himself in the position of ultimate active authority over the guards' behavior calls this into question. Many designers of such experiments would summarily throw out such a study based on this alone.

❖ Some researchers have also questioned why Zimbardo neglected the effect of individual personalities, instead generally attributing all behavior to the prison environment. How did John Wayne's behavior as a guard compare to his behavior outside the experiment? Was he generally a friendly guy, or might he already have been a royal jerk? We don't know, so there was insufficient data to conclude that his behavior was changed by the experiment.

❖ The statistical validity of the sample of participants, 24 male Stanford students of about the same age, has been called into question as being too small and restrictive to be generally applicable to the population at large.

❖ I have one other issue with Zimbardo's results that I didn't find anyone else raising. Zimbardo has dedicated much of his career to the promotion of the idea that bad environments drive bad behavior. I tend to be cautious of claims coming from sources dedicated to promoting them. The scientific method starts with a null hypothesis, not with a preconceived notion to justify; and that process invariably produces data that do not support the conclusion, and theories tend to change over time as a result. By my analysis, Zimbardo appears to be cherrypicking his results to justify the same conclusion that he has been promoting throughout his career. This doesn't make him wrong, it just gives me cause for skepticism.

❖ Finally, It's worth mentioning that by today's standards, the Stanford Prison Experiment was unethical and could never be performed in the United States. However, this point is not relevant to the validity of the results, and in any event, it was perfectly legal at the time.

Dr. Zimbardo and the Stanford Experiment came into the news again in 2004, following the Abu Ghraib prison scandal in Iraq. American prison guards were accused of cruelty to Iraqi prisoners — the great Naked Human Pyramidgate scandal. A number of soldiers and senior officers were court martialed and imprisoned or demoted. The prosecutors claimed that "a few bad apples" were responsible. The defense disagreed, and called in Dr. Zimbardo as an expert witness to testify that it was the environment that was responsible, not the individuals. "You can't be a sweet cucumber in a vinegar barrel," he famously said. The court disagreed, finding (rightly, as many would say) that individuals must be held accountable for their own actions, and the few bad apples went to jail. Dr. Zimbardo then wrote the book *The Lucifer Effect*, drawing further parallels between his prison experiment and the Abu Ghraib scandal.

Psychology is complicated, and there will probably never be a perfect theory explaining all human behavior; so people should never assign too much significance to the results of any given experiment like the Stanford Prison Experiment. And, when an experiment receives a large amount of scholarly criticism from mainstream science, as this one did, you have very good reason to look past its portrayal in the popular media and, instead, be skeptical.

48. Should You Take Your Vitamins?

What do you do when someone in your household catches a cold? If you're like me, you run for the medicine cabinet and start megadosing yourself with vitamin C. Everyone knows this is the best way to stave off a cold. Everyone also seems to think that vitamins are like spinach for Popeye: Whether you need them or not, taking extra vitamins when you don't necessarily have a deficiency is perceived to bestow some sort of superhealth, as if you can somehow be even healthier than healthy. If you take any vitamin supplement, and you do not have a diagnosed vitamin deficiency, please read this chapter carefully.

The idea that vitamin C is a wonder drug for preventing colds and other illnesses became widely popular around 1970, due largely to some books written by one of our greatest scientists, Linus Pauling. Along with Marie Curie, Pauling shares the distinction of being one of only two people to receive Nobel Prizes in two different fields. He received his first in chemistry in 1954 for his pioneering work characterizing the nature of chemical bonds. He received the Peace Prize in 1962 for his work warning of the dangers of nuclear weapons proliferation and radioactive fallout. Ironically, it was this type of activism that had almost prevented his being able to travel to Stockholm to accept his first prize: In those days, the United States was in the habit of denying passports to citizens who were sufficiently outspoken against nuclear weapons. Pauling's contributions to 20th century science are inestimable. He described the structure of the atomic nucleus. He was a key player in the theorized, and later proven, helical structure of DNA. He was practically the founding father of the whole science of molecular biology. He was even involved in the development of one of the first commercially available electric cars, the 1959 Henney Kilowatt.

So how is it possible that such an accomplished genius could be fundamentally wrong for much of his career? Let's set aside the notion that a scientifically brilliant mind must always be a rational, objective, and unbiased mind, and look at Pauling's later writings. In 1970, he published a book called *Vitamin C and the Common Cold*, in which he outlined the concept he called orthomolecular medicine. Orthomolecular means the "right molecule". Its central thesis is that megadoses of vitamin C prevent colds and can prevent or treat other conditions. He expanded this to include cancer when he wrote 1979's *Vitamin C and Cancer*. In 1986 he published *How to Feel Better and Live Longer*, in which he broadened all his concepts and claimed that megadoses of all vitamins would improve your overall health, would slow aging, and increase your enjoyment of life. It's important to note that orthomolecular medicine is not a concept shared by responsible doctors or nutritionists; it is squarely in the alternative medicine camp. The Mayo Clinic tested Pauling's claims about treating cancer with vitamin C in three different randomized controlled trials, all of which showed no beneficial effect. Pauling spent years passionately trying in vain to discredit these trials, which created something of a gap between himself and mainstream medical science. In fact, despite the staggering importance of his earlier contributions to science, by the end of his life Dr. Pauling was largely regarded as a crank by his former colleagues, much like Nikola Tesla, by some accounts.

There have been at least 30 well-performed controlled trials to find out whether vitamin C at various dosages can prevent colds, or reduce their severity. What these studies have determined is that vitamin C has no preventive value at all; you're just as likely to catch a cold if you take daily megadoses as you are if you take nothing at all. A few of these trials did find small reductions in the severity or duration of the colds, but most trials did not show even this small effect.

I know all of this, but every time the cold and flu season comes around, I still catch myself eyeing that vitamin C bottle. I'm like the jungle native who's been baptized by the missionaries, but whenever the volcano erupts, I still run to the stone pagan idol. My own experience is that I've never gotten sick whenever I've been regularly taking vitamin C. And, this is the same experience reported by a lot of people. So it would seem that our own experiences support what Dr. Pauling was saying, and disputes what testing has revealed. We all have to believe our own eyes, and to believe our own first-hand experiences, right?

Well, yes, but we also have to understand the way our brain interprets our experiences. Practically the whole reason for the science of psychology is that what we think and feel is not necessarily 100% translated to the real world. It's certainly possible to misinterpret something someone tells us, so isn't it also possible to misinterpret other experiences? Well, we do misinterpret our own perceptions and our own experiences, and we all do it every day. Is it possible that you did have a minor cold, but since you were taking vitamin C the thought never entered your mind that it could be a cold? Maybe you just attributed it to seasonal allergies, and even though the facts are that you got a cold while taking vitamin C, your own perception confirms that the vitamin C was 100% effective. Is it possible that you don't exactly remember the number of colds you got last year? Of course it is. And, perceptual biases aside, is there any chance that you wouldn't have happened to catch a cold anyway? Of course that's possible too.

Thus, you can't reasonably consider your own experience as evidence that vitamin C is effective against colds. Your only evidence is anecdotal and unreliable due to a variety of perceptual phenomena; and in any case your own test of vitamin C was an uncontrolled, unblinded test with no statistical validity.

But you're not alone. A lot of people believe that vitamins will prevent colds, and the alternative medicine industry has always been quick to capitalize on this. There's a product called Airborne that is a repackaging of an ancient Chinese remedy called *yin chiao*. It contains undisclosed quantities of a few herbs and vitamins, including vitamins A, C, and E. Their marketing slogan is "Invented by a schoolteacher", which for some reason people view as meritorious, even though all it really means is "Invented by someone with no medical background whatsoever". For a decade they made false advertising claims that their product could prevent and treat colds. Sure enough, eventually the law caught up with them, and fined them $23 million and ordered them to refund the purchase of anyone who ever bought their product. They're still in business, although they now make the claim that Airborne "boosts your immune system". As any doctor will tell you, "immune system boosting" is pure pseudoscience. It's medically meaningless, but that's a whole other subject.

Not only is there a lack of evidence that these products have any beneficial effects, there is well established evidence that they can be dangerous. If you take the recommended dose of Airborne to fight a cold, you're taking enough of a vitamin C overdose to put yourself at risk of kidney stones. Although the popular folk wisdom teaches that extra vitamins are simply excreted in the urine, this is largely untrue. Vitamin overdosing is called hypervitaminosis or vitamin toxicity, and can lead to serious effects. Hypervitaminosis A can lead to birth defects, liver problems, osteoporosis, skin problems, and hair loss. Hypervitaminosis D can cause deydration, vomiting, anorexia, hypercalcemia and kidney damage including kidney stones. Hypervitaminosis E can lead to blood problems including high cholesterol and can act as an anticoagulant. It should be noted that to be at risk of any of these conditions, you would need to significantly overdose over a long period of time. Brief or modest overdoses of any vitamin supplement are unlikely to cause problems. Interestingly, vitamin K is an effective treatment for many hypervitaminosis toxicity conditions.

A common criticism I receive is "Why should I believe you, when upstanding companies like Airborne, and practically everyone else in the world, tells me I should take vitamins?" Well, I hope you don't trust me. I hope that if you're truly interested, you'll ask a medical doctor. And by "doctor" I don't mean a naturopath, a health food store clerk, or anyone else who's in the business of selling you vitamin supplements. The simple fact is that nearly everyone who eats anything close to a balanced diet in any developed country is extremely unlikely to have a vitamin deficiency. Thus, there is no plausible benefit to vitamin supplementation for general health or wellness. There is no such thing as healthier than healthy.

49. When People Talk Backwards

Just when you thought there was nobody in the world crazier than yourself, along come people who believe that we all subconsciously say what we really mean in reverse, through the unconscious but deliberate choosing of careful words which, if played backwards, say what we actually mean. Get it? The idea is that I think some coffee is really horrible but I still want to be polite, my brain will subconsciously choose words to make my polite compliment that, if played backwards, would say: This coffee stinks.

Proponents of this hypothesis call it Reverse Speech, because they were really creatively inspired on the day they named it. This is a small group of people — I believe there were six of them at last count — who take this completely seriously and believe that a whole world of secret information and opportunities is waiting to be unlocked by analyzing peoples' speech in reverse. They turn first to world leaders, play their speeches backward, and listen to learn what they believe is the truth underlying the speech.

A leading advocate for reverse speech, also called backward masking, is David John Oates, an Australian. He's written several books on the subject and even used to have a syndicated radio show promoting his theory. Just about any time a reverse speech expert is interviewed on television, it's David John Oates. His web site is ReverseSpeech.com, and it's loaded with all the examples you could ever hope to hear, as well as quite a few products and services he'd like to sell you if you believe his claims. He believes strongly that the human brain secretly encodes its actual meaning in reverse into a person's normal speech. You can use this to your advantage in business, by decoding what the people across the table are actually telling you; and you can even use it in personal development by listening to your own speech backwards and learning more about what you really want. One of the examples from

ReverseSpeech.com is of a man giving a talk. He says "More energy and money and effort", and when you play it backwards, turns out he was trying to comfort you with the message "You're frightened, lean on me".

Pretty interesting, but not necessarily convincing to a skeptic. A skeptic is more likely to dismiss these guys as conspiracy nuts and laugh at what paranoid delusionals they are, but it's actually way cooler and more interesting (and more constructive) to ask if there is any science behind what they're claiming. I'm not talking about science supporting the claim that people say what they actually mean in reverse; I'm talking about science behind the perception of order from chaos. And, it turns out, there is good science behind it. The journal *Science* published an article in 1981 by Remez, Rubin, Pisoni, and Carrell called *Speech Perception Without Traditional Speech Cues*. By playing what they called a "three-tone sinusoidal replica", or a complicated sine wave sound, they found that people were able to perceive speech, when in fact there were no traditional speech sounds present in the signal. So rather than laughing at a reverse speech advocate, instead appreciate the fact that there is good science driving their perception of what they're hearing. They're not making anything up, they're just unaware of the natural explanation for their phenomenon.

This phenomenon is called pareidolia, which we talked about not too long ago when we discussed the face on Mars. Pareidolia is the perceptual phenomenon by which we perceive familiar patterns in disorder. It is the brain's incredible computing power that lets us recognize people, understand language, and read handwriting. For the brain to have this capability, it necessarily results in the ability to perceive patterns where none in fact exists. Most of us can say "Hey, that tree bark looks like Ernest Borgnine," without actually concluding that Ernest Borgnine has somehow become a tree. Our intelligence allows us to not make that mistake. But sometimes a horse might see a garden hose on the ground; its pareidolia tells it that it's a snake, but it lacks sufficient intelligence to overcome the instinctive recognition. I'm not saying that reverse speech believers lack intelligence, only that they lack critical thinking skills; because there is a genuine gray area where it's hard to tell if a pattern is accidental or deliberate. But speech is a deliberate speaking action, so the reverse speech advocates do have a point they can make. It's not an accident of nature like the tree bark, speech is the deliberate result of a speaker's brain communicating. What the reverse speech advocates are missing is that the well-known, well-understood, and well-evidenced phenomenon of pareidolia is a much more reasonable, simple, and probable explanation for why we can often perceive patterns in meaningless noise, in this case reverse speech.

This also fully explains a couple of other pop culture phenomena: Satanic messages encoded in rock music played backwards, and EVP, the electronic voice phenomena claimed by ghost hunters. There's a really popular clip of Jim Morrison of The Doors singing the line "Treasures there" from *Break On Through*. When you listen to it backwards, he says "I am Satan".

But, to get it to sound like "Satan", you have to be a little disingenuous with the razor blade. Turns out that if you don't edit it very precisely to isolate the word Satan, it sounds more like "Sata-Schnigel". So if reverse speech is real, Jim Morrison's true intention in life was to inform us that he's Sata-Schnigel. And, as you can probably surmise, even that very impressive "You're frightened, lean on me" is much less convincing without the incisive editing. It sounds more like "Lean on Me-journey Elm". This is what we in the brotherhood call "cherrypicking".

Recordings of alleged ghost voices, usually called Electronic Voice Phenomena, fall into three categories: First, hoaxes; second, undetermined; and third and most commonly, audio pareidoliac cases of mistaken identification. You hear some random anomalous sound on the tape, and your brain does its best to make sense of it, often turning it into speech. If the words that the ghost hunters claim are spoken are at all indistinct or ambiguous, there is a very probable explanation for them that's not "a ghost". You're hearing some sound, and unless you were present throughout the tape's entire history (which you probably weren't), it's some sound of unknown origin that, to your brain, sounds vaguely like speech, and isn't it interesting that it's always in the ghost hunter's own language and dialect? The human brain is hardwired to hear its own language in otherwise meaningless noise. If it wasn't, you'd never be able to recognize your own name when someone calls out to you in a noisy room full of people.

Whether that noise is human speech played backwards, music played backwards, traffic sounds, or random noises in a graveyard or haunted house, your human brain will process it and find intelligible speech. It's the way your brain works, it's not evidence of ghosts, Satanic messages, and certainly not of something as childish as reverse speech.

50. KING TUT'S CURSE!

We've all heard the story of the mummy's curse, and we've all heard the popular explanation — but you may not know the numbers behind the story. Today we're going to point our skeptical eye at King Tut's curse, and find out exactly how the story goes, explore the scientific-sounding explanation proposed by the media, and finally, we're going to look at what really happened.

In 1922 Howard Carter was exploring the Valley of the Kings in Egypt, with his friend and financial backer George Herbert, the 5th Earl of Carnarvon. After a 15 year search, they opened the fabulous tomb of King Tutankhamun, the most spectacular tomb found to date, and now known as KV62. Carter poked a hole through the seal and peered inside, and when Lord Carnarvon asked if he could see anything, Carter famously replied "Yes, wonderful things."

But things got less wonderful rather quickly, so the story goes. Front and center in the antechamber was a clay tablet, deciphered by one of Carter's colleagues, that read:

> *Death will slay with his wings whoever disturbs the peace of the pharaoh.*

None of them seemed too concerned about Death's wings, because the archaeologists immediately proceeded to apply their early 20th century archaeological ignorance to King Tut and his goodies. Items were collected and broken, and even King Tut's mummy itself was said to have been chopped into pieces and set out in the sun where it quickly deteriorated. According to legend, Lord Carnarvon soon died from a mosquito bite; and simultaneously, his three-legged dog howled and dropped dead, and all the lights in the city of Cairo suddenly went out.

And then others of Carter's party began to die of mysterious causes. In fact, more than two dozen men were said to have fallen to the curse. Carter himself, it appears, had to suffer the fate of watching all his friends and associates drop off like flies all around him. Even his beloved pet canary was killed by a cobra in a freak incident. The newspapers trumpeted the terrors of the mummy's curse to all the world. Carter bore these miseries until he finally died himself, sixteen years after unleashing the curse.

Now the critical mind can easily find many causes for skepticism with this story:

❖ First, the accounts of the curse all come from 1920's-era newspapers, well known for sensationalism and expansion of facts to make great headlines. Reliable records of what happened to Carter's people after they left the dig are hard to come by. References to the simultaneous death of dogs, the canary, and the lights going out in Cairo are found only in these unreliable newspaper reports and so can be considered anecdotal at best. Anyway, how reliable were the public utilities in Cairo in 1920? Would it be more or less remarkable if the lights *hadn't* gone out at any particular time?

❖ Second, Lord Carnarvon was known to be in pretty frail health at the time, and infection was a common cause of death. He had aggravated the mosquito bite on his cheek while shaving, and developed erysipelas resulting in septicemia and pneumonia. There was no curse needed to explain the dangers of these conditions.

❖ Third, the explanation that the curse's effect on Carter himself was to leave him alive and well while others died is clearly a post-hoc rationalization. Sure, I suppose it's possible that Carter's long healthy life could be evidence of a curse, but the lack of an effective curse is probably a better explanation for it.

❖ Fourth, and this goes back again to the pulp-fiction nature of the newspapers of the era, is the inconsistencies among various versions of the tale, notably the alleged stone tablet bearing the curse. It should be noted that there is no record of any written curse, either in Carter's own documents or in any modern collections; at least not associated with Tutankhamun's tomb specifically. Another post-hoc rationalization exists to explain the absence of a written record: It was expunged to avoid frightening the locals. Again, a better explanation is that such a written curse did not exist.

One of the first people to present a serious scientific explanation for the deaths associated with King Tut's curse was Dr. Caroline Stenger-Phillip, who proposed in 1986 that ancient mold in the tomb could have caused potentially fatal allergic reactions. Since fruits and vegetables and other organic items were buried in tombs, and since the tombs were completely hermetically sealed, it is plausible that mold spores could have existed and remained viable through the millennia.

This proposal has become known as "tomb toxins", and has been broadened to include other compounds, such as two molds that are found on ancient mummies, *Aspergillus niger* and *Aspergillus flavus*, and that can be potentially harmful to people with weakened immune systems. Bacteria are also found in tombs, including *Pseudomonas* and *Staphylococcus*. And don't forget the chemicals used in embalming the mummies: ammonia, formaldehyde, and hydrogen sulfide. Tomb toxins do sound like a plausible explanation for the mummy's curse. Most of us have heard this explanation at one time or another and thought "Ah, that explains it quite neatly."

But unfortunately, tomb toxins do not explain the deaths from Carter's group very well at all. Even in the unlikely event that members of Carter's party received lethal doses of any or all of the above, such death would have followed quite quickly; it wouldn't have been delayed by the months or years reported among the victims of King Tut's curse. Even Lord Carnarvon's death, the one most closely associated with the curse, occurred six months after he entered the tomb.

Another problem with the tomb toxin explanation is that it sounds good to a layperson, but it is, in fact, armchair science. It's a reasonably plausible idea, but one that has never actually happened in the real world. *National Geographic* is among those who have delved into this subject in detail, and found that working Egyptologists are not concerned about the possibility of tomb toxins. They've never heard of any colleagues suffering from it; thousands of tourists go in and out of the tombs every day with no ill effects, and even when Egyptologists do wear masks during excavations it's because of dust, not tomb toxins. F. DeWolfe Miller, professor of epidemiology at the University of Hawaii said "Given the sanitary conditions of the time in general, and those within Egypt in particular, Lord Carnarvon would likely have been safer inside the tomb than outside."

So we have two things at this point in our investigation: First, really weak and primarily anecdotal evidence that anything unusual happened; and second, a hypothesized cause that turns out to be quite a poor fit for the observed data. King Tut's curse is beginning to look about as withered up as he looks himself.

In 2002, the *British Medical Journal* published a study by Dr. Mark Nelson from Monash University in Australia. He decided to take a statistical look at the people who were actually there, and see if their dates of death actually were accelerated as a result of exposure to any possible curse. He performed a retrospective cohort study, which is a specific type of analysis based on medical records of certain groups of people. Nelson considered only the Westerners in Carter's party, since there was a difference in life expectancy between Westerners and Egyptians. He defined "exposure to the curse" as participation in any of four specific events where sacred seals were breached in the tomb, the sarcophagus, and the mummy itself. And then the number crunching began.

To better understand these results, it's necessary to comprehend what's meant by a "p-value". It's a term used by statisticians, and although the actual definition is complex and hard to grasp, a "good enough" layman's definition for our purposes is the probability that your test results are consistent with expected random variations. A p-value of 0, the lowest possible, means there's a 0% chance that your test results are due to normal random variances, so low p-values generally mean that your results are significant. A p-value of 1, the highest possible, means that your results are 100% consistent with what we'd expect to see from normal random variations, therefore your results are quite probably insignificant.

Of 44 Westerners present, 25 were exposed to the curse. Those 25 lived to an average age of 70, while those not exposed lived to 75. The p-value of this difference was .87, so there's an 87% chance that this difference was merely due to chance. Average survival after the date of exposure was 20.8 years for the exposed group, and 28.9 years for the unexposed group. While this sounds like a large difference, the p-value was .95, meaning there's a 95% chance that you'd have such a difference anyway due to random variation. Nelson's conclusion: "There was no significant association between exposure to the mummy's curse and survival and thus no evidence to support the existence of a mummy's curse."

So, we end up with one piece of hard, testable evidence: Statistically speaking, nothing unusual happened in the Valley of the Kings; but pop culture gained one more rich layer of adventure fiction.

ABOUT THE AUTHOR

Brian Dunning is an author, podcaster, speaker, and software investment professional in Southern California. It is all a desperate flurry of activity to avoid anything that smacks of real work.

His free weekly podcast *Skeptoid: Critical Analysis of Pop Phenomena* began in 2006, taking the underdog side of science and rationality amid the overwhelming majority of noise in the media promoting pseudoscience, alternative anything, and unconditional acceptance of the paranormal. The choice between pseudoscience and science is the choice between the developmental stagnation of the Dark Ages, and progress.

Brian takes every opportunity to speak on critical thinking issues at colleges and local groups. In his spare time he enjoys Jeeping in the desert with his family and playing beach volleyball (badly).